TABLE OF CONTENTS

Recreation Area 1

Durango, Silverton, Bayfield

Trail Name	Trail No.	Map No.	Page
Calico Nat. Recreation	208	1	7
Fish Creek	647	1	7
Geyser Spring	648	1	7
Ice Lake	505	2	9
Rico-Silverton	507	2&6	9
Cross Mountain	637	2	9
East Fork	638	2&6	9
Cunningham Gulch	502	3	10
Columbine Lake	509	2&3	11
Highland Mary	606	3	11
Whitehead	674	3	11
Continental Divide	813	3&7	11
Loading Pen	738	4&5	13
Wildcat	207	5	15
Rough Canyon	435	5	15
Bear Creek	607a	5	15
Gold Run	618	5	16
Horse Creek	626	5	16
Johnny Bull	639	5	16
Tenderfoot	644	5	16
Priest Gulch	645	5	17
Calico	649	5	17
Ryman Creek	735	5	17
Twin Spring	739	5	17
Pass Trail	500	6	19
Graysill	506	6	19
Engineer Mountain	508	6	19
Cascade Creek	510	6	19
Elbert Creek	512	6	20
Little Elk Creek	515	6&10	20
Dutch Creek	516	6&10	20
Goulding Creek	517	6	20
Big Bend	519	5&6	21
Corral Draw	521	6	21
Salt Creek	559	5&6	21
Engine Creek	657	6	22
Spud Lake	661	6	22
Animas River	675	6&7	22
Coal Creek	677	6	22
Deer Creek	678	6	23
West Lime Creek	679	2,6&7	23
Elk Creek	503	7	25
Needle Crk/Chicago Basin	504	7	25
Crater Lake	623	7	25
Molas	665	7	25
Kennebec Pass-Junction Crk. Rd.	553	9&10	27
West Mancos	565	9	27
Highline Loop Nat. Rec.	607b	5&9	27
Sharkstooth	607c	9	27
Hermosa Creek	514	6&10	29
Jones Creek	518	6&10	29

Trail Name	Trail No.	Map No.	Page
South Fork	549	6&10	29
Clear Creek	550	10	29
Haflin Creek	557	10	30
Sliderock	622	9&10	30
First Fork	727	10&11	30
East Creek	535	11,Area 2, M4	33
Youngs Canyon	546a	11	33
Shearer Creek	558	11	33
North Canyon	656	11	33
Lost Lake	663	11	33
Red Creek	726	11	34
Lake Eileen	668	11	34
Vallecito Reservoir			*35*
Colorado Trail			*37*
A-B	1776	7	44
B-C	1776	2,3,6&7	44
C-D	1776	5,6&9	44
D-E	1776	10	45
Loop	1776	2,3,6,&7	45

Recreation Area 2

Pagosa Springs, Weminuche and South San Juan Wilderness

	Trail No.	Map No.	Page
Continental Divide Trail			*47*
A-B	813		51
B-C	813		51
C-D	813		52
D-E	813		52
E-F	813		53
F-G	813		53
Williams Creek	587	2	66
Piedra River	596	4&5	66
Fourmile Falls	579	5	66
Little Blanco	572	6&7	66

Recreation Area 3

Dolores, McPhee Recreation Area *69*

Map Symbol Explanation 13
Forest Service and Other Phone Numbers 31
Forest Service Trail Rating Method 45
Recommended Trail Activity Summary *67*

ISBN 0-930657-13-6

HOW TO USE THIS GUIDE	4
RECREATION AREA 2	47
RECREATION AREA 3	69

HOW TO USE THIS GUIDE

In this guide the San Juan National Forest has been divided into three recreation areas. Recreation Area 1 is the central section and can be accessed from Dolores, Mancos, Durango, Silverton and Bayfield. This section of the guide contains the most trails. Recreation Area 2 is the eastern area of the forest showing the Weminuche and South San Juan Wilderness areas. Section 2 is accessed from Pagosa Springs. Recreation Area 3 is McPhee Recreation Area located near Dolores. The numbered recreation map located in front of each area locates trails, campgrounds, rivers, lakes, wilderness boundaries and other physical features. Map scale is approximately 1"= 2 miles. Campgrounds are shown on the map with a number and a campground symbol. The campground information is located in each area and referenced by the map number.

Trail "quick reference" information bar heading is provided for each trail.
BAR HEADING EXPLANATION

Forest Service Trail number.

Trail Name

ONE-WAY distance of trail.

Beginning and ending trail elevations.

Ranger District that manages trail

Trail No.	Trail Name	Map Loc.	Distance	Difficulty	Beginning Elev.	Ending Elev.	Ranger District
712	Cottonwood	J 7	1.6 mi	Moderate	10,800'	10,600'	Pagosa

Map coordinates of trail location on San Juan National Forest map.

Forest Service trail rating. (See page 45)

TEXT EXPLANATION

ACCESS: Directions to reach trailhead.

ATTRACTIONS: Describes terraine, use and conditions of trail.

NARRATIVE: Describes scenery, destination, and wildlife that one might encounter.

USE: How many people use the trail. Catagories: Heavy, medium and light.

ACTIVITIES: Suggested recreation activities. NOTE: Trails may or may not be restricted to activities shown - call first.

USGS: Name of U.S. Geological Survey 7 1/2' topographic quadrangle on which the trail is shown. Scale 1" = 2000 feet.

NOTE: Information contained in this guide is for general reference only. Use good judgement when planning your trip. Forest Service difficulty trail ratings and directions are general guides, physical condition, age, weather, altitude and experience are factors that should be considered when planning your trip. If you have questions about conditions or trail routes contact the local Forest Ranger. The intended use of this guide is for trip planning only. Outdoor Books & Maps, Inc. is not responsible for mishap or injury from other than its intended use.

RANGER DISTRICT INFORMATION

Mancos/Dolores Ranger District
100 North 6th
Dolores, CO 81323
(970) 882-7296

Columbine Ranger District
367 S. Pearl Street
PO Box 439
Bayfield, CO 81222
(970) 884-2512

Pagosa Ranger District
2nd & Pagosa
PO Box 310
Pagosa Springs CO 81147
(970) 264-2268

Associate Forest Service Supervisor
701 Camino del Rio
Room 301
Durango, CO 81301

MAP SOURCES
San Juan National Forest Service Maps are sold at Forest Service offices, sporting goods and map stores. All you need for a trouble free trip is a Forest Service map, this Guide, and a U.S. Geologic Topographic map showing the trail. This combination of information and maps is unbeatable!

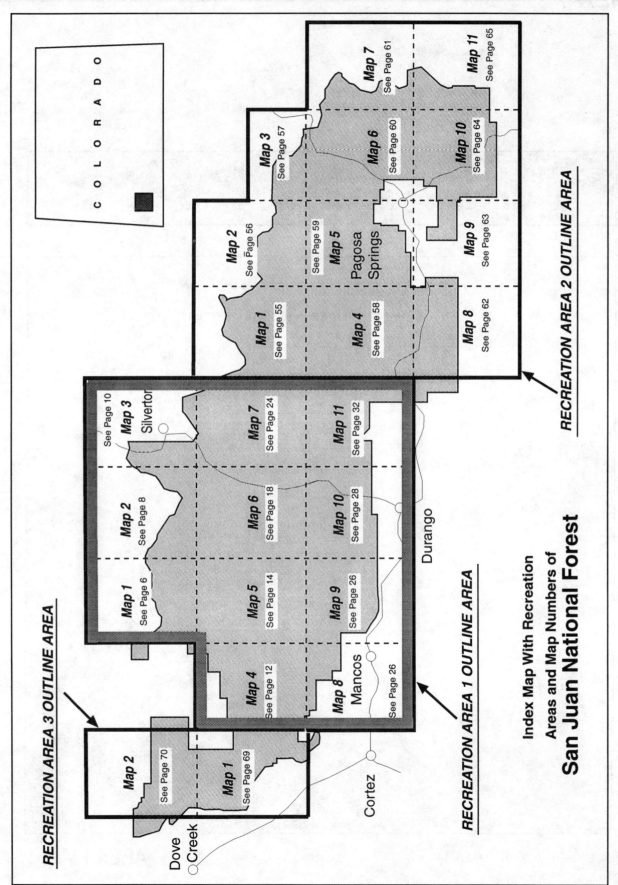

Index Map With Recreation
Areas and Map Numbers of
San Juan National Forest

Map No.	Name	Fee	No. of Units	Max. Length	Elev.	Toilets	Water	Ranger District
1.	Burro Bridge	$	15	35'	9,000'	Yes	Yes	Mancos/Dolores

CAMPGROUNDS LOCATED IN AREA 1 MAP 1.

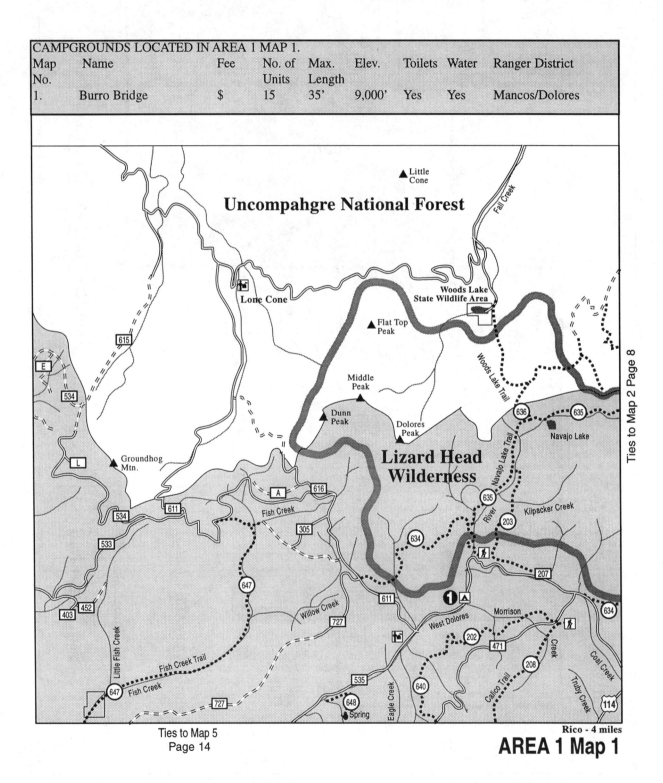

Trail No.	Trail Name	Map Loc.	Distance	Difficulty	Beginning Elev.	Ending Elev.	Ranger District
208	**Calico Nat. Rec.**	H 2	6.0 mi.	Difficult	10,150'	10,200'	Mancos/Dolores

ACCESS: From the "Meadows" located on FDR 535, seven miles above Dunton, take FDR 471. The Calico trailhead is approximately 1 mile south of this intersection. The trailhead has a gravel parking area, signing, hitching racks, and a restroom. The trail begins if you continue along FDR 471 to the end you will come to the southern end of the NRT trail at the west fork of Fall Creek. **NARRATIVE:** The Calico National Recreation Trail provides day use or an extended trail experience. It begins in an area of vast mountain meadows interspersed with woods of Englemann spruce and alpine fir. To the north rises 14,000 foot high El Diente Peak, which is within the Lizard Head Wilderness. The trail then moves south through wet, lush meadows and moves into a forest of spruce and fir as it follows the divide between the Dolores and West Dolores Rivers. It is climbing at an average grade of 8% to 11,866 foot high Papoose Peak. Before the trail reaches Papoose Peak, it has risen above timberline into a subalpine ecosystem. At the headwaters of the East Fork Fall Creek, the Calico NRT drops in elevation and returns into the spruce-fir forest or you can continue on to the junction with the west fork of Fall Creek which also drops in elevation and returns into the spruce-fir forest. The alpine ecosystem above 11,000 feet, is basically devoid of trees. Dominant plant species include sedges, willows, fescues and bluegrasses and numerous forbs. Permanent animal residents include the pika, yellow-bellied marmot, weasels and white-tailed ptarmigan. The deer and elk utilize this habitat during the summer. Migrant birds include the water pipit, rock wren, Wilson's warbler, and white-crowned sparrow. The Engelmann spruce, subalpine fir ecosystem associated plants include quaking aspen, Douglas fir, Rocky Mountain maple, common juniper, sedges, grouse whortleberry, buffaloberry, and heartleaf arnica. There are over 28 species of birds, including the golden eagle, that nest in this ecosystem. And over 24 species of mammals, from black bear and elk to the vagrant shrew. The Calico National Recreation Trail is a 14 mile loop formed from a section of the Calico Trail, East Fork Fall Creek Trail, West Fork Fall Creek Trail, and Winter Trail. Shorter loops can be made by hiking either the Calico portion of the loop and shorten the hike by coming down either the East Fork or West Fork of Fall Creek. Likewise a shorter loop is found using only the Winter and Fall Creek Trails. **USE:** Low. **ACTIVITIES:** HIKING. **USGS:** Rico, Dolores Peak Quads. **MAPS:** 1&5.

Trail No.	Trail Name	Map Loc.	Distance	Difficulty	Beginning Elev.	Ending Elev.	Ranger District
647	**Fish Creek**	G 2	7.5 Mi.	Difficult.	8,200"	9,520'	Mancos/Dolores

ACCESS #1: Fish Creek State Wildlife Area. Travel 13 miles north of Dolores on Hwy 145. Turn left on West Dolores Road (FDR 535), go approximately12 miles. Look for gravel road on left, approximately 1/4 mile before the large new West Dolores Road Bridge over Fish Creek. Take this gravel road (FDR 726) turning left before Dolores County Shop. The trailhead is at the end of road. Use 4 wheel drive in wet weather. **ACCESS #2:** Travel 13 miles north of Dolores on Colo. Hwy. 145. Turn left on West Dolores Road (FDR 535), go about 22 miles, turn left on FDR 611. Go about 3 miles, turn left on FDR 536. Use 4 wheel drive beyond this point. Continue for 2 miles, crossing Fish Creek near the foot bridge, keeping left at the fork. Trail sign is on the right before you cross the creek again. **NARRATIVE:** Trail follows Fish Creek, a major tributary of the West Dolores River through forests and open meadows. Motorized vehicles are not allowed. A 1,550' stretch of talus slope about 3 miles up from Fish Creek State wildlife Area can be difficult for horses. Stream crossing are 4-6 feet wide, 2 feet deep. Fishing is good. **USE:** Light. **ACTIVITIES:** HIKING, HORSES, FISHING. **USGS:** Groundhog Mountain, Clyde Lake Quads. **MAPS:** 1&5.

Trail No.	Trail Name	Map Loc.	Distance	Difficulty	Beginning Elev.	Ending Elev.	Ranger District
648	**Geyser Spring**	H 2	1. 25 Mi.	Easy/Mod.	8,600'	9,120'	Mancos/Dolores

ACCESS: The trail starts by crossing the West Dolores River approximately 2 miles below Dunton on FDR 535 (Dunton Road). This is approximately 17miles from the junction of Colorado State Hwy. 145 and FDR 535. The spot is at a small pull out off FDR 535 marked by a trail sign. However, the trail marker is repeatedly pulled down. Crossing the West Dolores River can be difficult, the water rushes by rapidly. A walking staff might be helpful. **NARRATIVE:** This trail ends up at a small hot pool fed by only true geyser in the state of Colorado. Although the frequency of the eruptions varies, 30 to 40 minutes intervals are most common. The action is slight and boils 12 to 15 minutes like a whirlpool, emitting strong sulfur gasses and rising only 12 to 15 inches above the motionless part of the spring. The temperature of the spring is 28°C., 82.4°F. This is not the ideal temperature for soaking, it's somewhat cool. The rocks lining the pools crude walls can be rough and abrasive. The walk up is gradual and passes through forests and small meadows. Evidence of mining activity is very apparent in the area. Sulfurous gases can be smelled along the upper trail. **USE:** Low. **ACTIVITIES:** HIKING. **USGS:** Dolores Peak, Groundhog Mtn., Clyde Lake, Rico Quads. **MAP:** 1

Map No.	Name	Fee	No. of Units	Max. Length	Elev.	Toilets	Water	Ranger District
1.	Sunshine	$	15	--	9,500'	Yes	Yes	Norwood
2.	Cayton	$	27	35'	9,400'	Yes	Yes	Mancos/Dolores
3.	South Mineral	$	26	45'	10,000'	Yes	Yes	Columbine

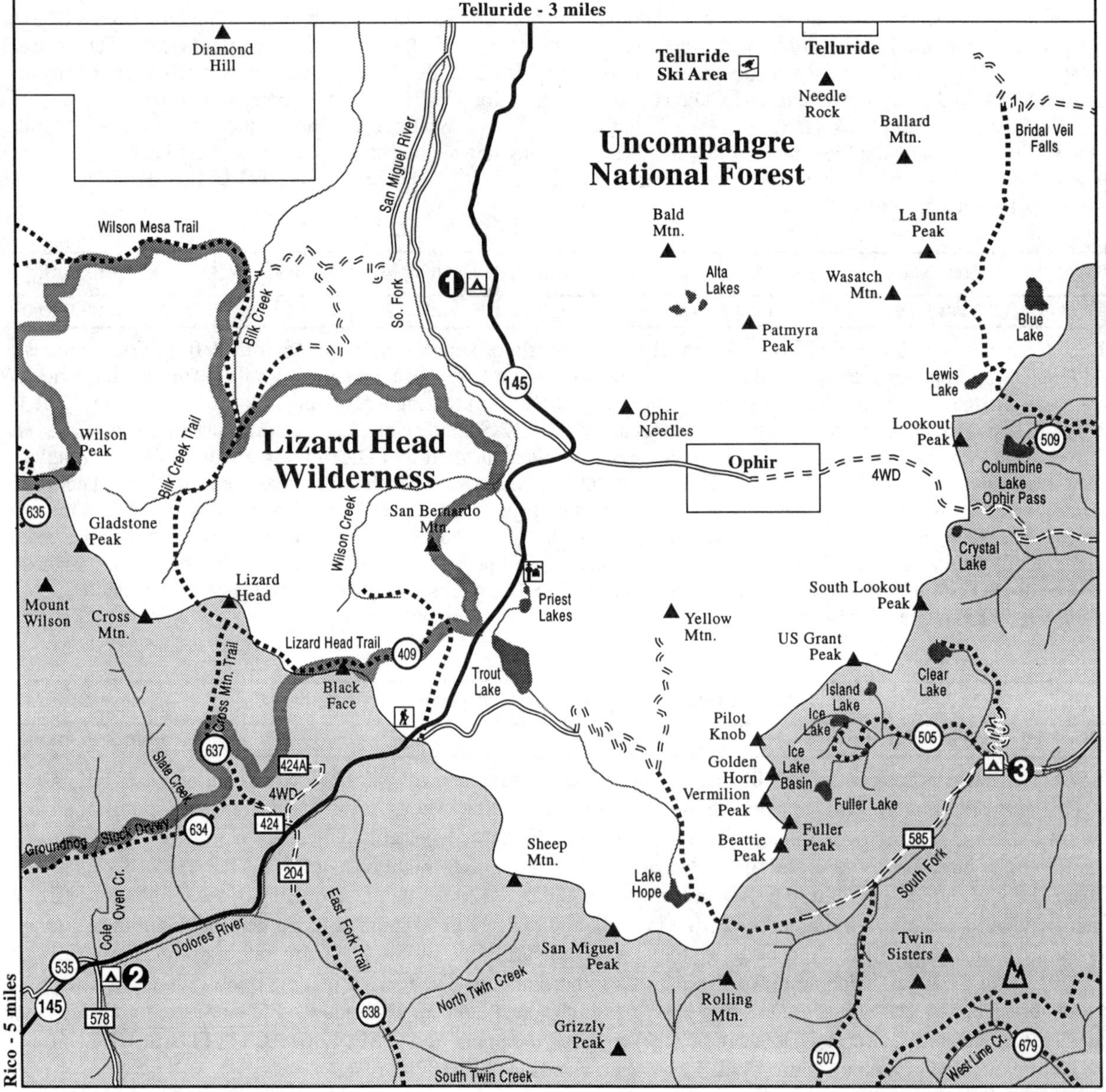

CAMPGROUNDS LOCATED IN AREA 1 MAP 2.

Telluride - 3 miles

Diamond Hill

Telluride Ski Area

Telluride

Needle Rock

Ballard Mtn.

Bridal Veil Falls

Uncompahgre National Forest

San Miguel River

Wilson Mesa Trail

Bilk Creek

So. Fork

Bald Mtn.

La Junta Peak

Alta Lakes

Wasatch Mtn.

Blue Lake

Ties to Map 1 Page 6

Ties to Map 3 Page 10

145

Patmyra Peak

Lewis Lake

Wilson Peak

Bilk Creek Trail

Lizard Head Wilderness

Ophir Needles

Lookout Peak

509

635

Gladstone Peak

Wilson Creek

San Bernardo Mtn.

Ophir

4WD

Columbine Lake Ophir Pass

Crystal Lake

Mount Wilson

Cross Mtn.

Lizard Head

Priest Lakes

South Lookout Peak

Yellow Mtn.

US Grant Peak

Clear Lake

Cross Mtn. Trail

Lizard Head Trail

409

Trout Lake

Pilot Knob

Island Lake

Ice Lake

505

637

Black Face

Golden Horn

Ice Lake Basin

3

Slate Creek

424A

4WD

634

424

Vermilion Peak

Fuller Lake

Groundhog Stock Drwy.

204

Sheep Mtn.

Beattie Peak

Fuller Peak

585

South Fork

Cole Oven Cr.

Dolores River

East Fork Trail

Lake Hope

535

145

578

San Miguel Peak

Twin Sisters

638

North Twin Creek

Rolling Mtn.

507

679

West Lime Cr.

Rico - 5 miles

2

Grizzly Peak

South Twin Creek

Ties to Map 6 Page 18

AREA 1 Map 2

Trail No.	Trail Name	Map Loc.	Distance	Difficulty	Beginning Elev.	Ending Elev.	Ranger District
505	**Ice Lake**	K 2	4.5 Mi.	More Diff.	9,840'	12,260'	Columbine

ACCESS: Follow U.S. 550 north of Silverton for approx. 2 miles and turn onto FDR 585 (South Mineral Creek). Limited parking is available. **ATTRACTIONS:** This trail is short and steep. It is uphill all the way to the lake basin. The first half of the trail is located below timberline and passes through aspen and conifer vegetative zones. Lower Ice Lake is located slightly below timberline at the base of a towering ridgeline. The upper half of the trail switchbacks up a cliff and then levels out when it reaches the basin. The basin is typically covered by wildflowers in late July and in August. This lake basin is surrounded by several peaks which include Grant Peak, Pilot Knob, the Golden Horn, Vermilion Peak, Fuller Peak, and Beatie Peak. As with all areas above timberline, there is little natural shelter from the elements. Storms can move into these areas very rapidly. All water in this area should be filtered for Giardia. **USE:** Heavy. **ACTIVITIES:** HIKING. **USGS:** Ophir Quad. **MAP:** 2.

Trail No.	Trail Name	Map Loc.	Distance	Difficulty	Beginning Elev.	Ending Elev.	Ranger District
507	**Rico-Silverton**	K 2	8 Mi.	More Diff.	10,800'	11,200'	Columbine

ACCESS #1: Take U.S. 550 north of Silverton for approx. 2 miles and turn onto the South Mineral Rd. (FDR 585) Follow the dirt road past the campground for approx. 2 miles to the Bandora Mine. Turn left slightly before the mine and follow the road across the creek. 2WD with good clearance should be adequate. Limited parking is available. **ACCESS #2:** Take U.S. 550 30 miles north of Durango to FDR 578 and go 16 miles on this road until the trailhead is reached. Limited parking is available. 4WD or strong 2WD with good clearance may be required to reach this trailhead located at the top of Bolam Pass. **ATTRACTIONS:** This trail ascends two high mountain passes, thus affording excellent views and good opportunities for photography. Grizzly Peak, Rolling Mtn., and Graysill Mtn. are all accessible from the Rico-Silverton Trail providing mountain climbing possibilities for those interested. The trail rises above timberline on two occasions, but for the most part winds through a subalpine coniferous forest. The alpine meadows above timberline typically abound with wildflowers at the end of July and in August. Water and campsites are easily found in this area. Water in this area should be filtered to prevent problems with Giardia. **USE:** Medium. **ACTIVITIES:** HIKING. **USGS:** Ophir, Engineer Mountain, Hermosa Peak Quads. **MAPS:** 2&6.

Trail No.	Trail Name	Map Loc.	Distance	Difficulty	Beginning Elev.	Ending Elev.	Ranger District
637	**Cross Mountain**	J 2	2.7 Mi.	More Diff.	10,200'	12,200'	Mancos/Dolores

ACCESS: Continue north along Colorado State Hwy. 145, 10 miles beyond Rico, Colorado. Turn west on FDR 424. Cross the stream and continue for approximately 1/4 mile further to a bulletin board which marks the trailhead. A few minutes up the trail will bring you to a "Y" in the trail. This is the intersection with part of the groundhog stock driveway. Stay right for the Cross Mountain Trail **NARRATIVE:** Cross Mountain Trail provides access to the Lizard Head Wilderness area. It is important to remember that within the wilderness certain special regulations apply. No motorized or mechanized devices are allowed. This includes motorbikes, mountain bikes and chain saws. Upon leaving a wilderness campsite every effort should be made to renaturalize the area. Cross Mountain Trail is open to horses and vigorous hikers who don't mind the more difficult climb and elevation. As you begin this hike you will pass the remains of an old sawmill. After reaching the crest, views of Lizard Head, Mt. Wilson, El Diente, Sheep and San Juan and Rico Mountains are quite impressive. A wonderful array of wildflowers can to generally viewed among the mountain meadows, Including masses of columbine and alpine sunflowers. This trail provides access to the more popularly used Lizard Head Trail and could be used to make a partial loop trip with it. The first 1-1/2 miles of trail follows an old road which has been obliterated. It will be years before natural vegetation takes hold and problems with erosion are eliminated. **USE:** Low Moderate. **ACTIVITIES:** HIKING, HORSES. **USGS:** Mt. Wilson Quad. **MAP:** 2.

Trail No.	Trail Name	Map Loc.	Distance	Difficulty	Beginning Elev.	Ending Elev.	Ranger District
638	**East Fork**	J 2	7.5 Mi.	More Diff.	10,120'	11,120'	Mancos/Dolores

ACCESS #1: Take Hwy 145 approx. 9 miles north of Rico or 2 miles south of Lizard Head Pass to FDR 204. Take FDR 204 approx. 1/2 mile to the Trailhead located next to old cabin. 4 WD recommended due to poor condition of road. **ACCESS #2:** Take Hwy 145 approx. 6 miles north of Rico to The Hermosa Pass Road (FDR 578) also the turnoff to Cayton Campground. Take FDR 578 up over Bolan Pass to its intersection with FDR 578B at Graysill Lake (4WD Recommended). Take FDR 578B approx. 1.3 miles until you find a trail sign on your left. Or you can take FDR 578B to its end (4WD required) and pick up there. **ATTRACTIONS:** This trail follows the canyon carved by the East Fork of the Dolores River. It slowly ascends a spruce-fir and aspen forest. Trout Fishing available in the East Fork. **USE:** Medium. **ACTIVITIES:** HIKING, FISHING, MTN. BIKING. **USGS:** Mount Wilson, Hermosa Peak Quads. **MAPS:** 2&6.

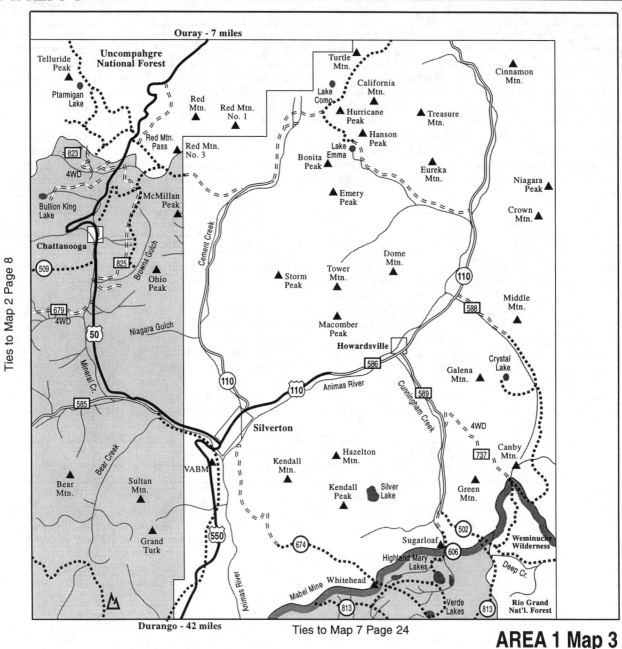

Ties to Map 2 Page 8

AREA 1 Map 3

Ties to Map 7 Page 24

Trail No.	Trail Name	Map Loc.	Distance	Difficulty	Beginning Elev.	Ending Elev.	Ranger District
502	**Cunningham Gulch**	M 2	2.5 Mi.	Most Diff.	11,400'	12,200'	Columbine

ACCESS : Follow U.S. 550 north to Silverton. Continue north through the town of Silverton on FDR 586 toward Howardsville. Approx. 3 miles turn onto the Cunningham Gulch Road, FDR 589 and follow it to the trailhead. 2WD vehicles may not be able to drive the last 3/4 mile of the road. **ATTRACTIONS:** This trail is the shortest route available to the Continental Divide on this district. It provides excellent opportunities for rock-hunting, fishing, and observing high country flora and fauna. August is the best month to observe wildflowers blooming. The Continental Divide provides excellent views and opportunities for photography especially of the Grenadier Mountains which are located south of the Divide Trail. Most of this hike is above timberline and acclimation to the elevation should be taken into account. Water availability should not be a problem, but all water should be filtered to avoid any problems. **USE:** Medium. **ACTIVITIES:** HIKING, FISHING. **USGS:** Howardsville Quad. **MAP:** 3.

NO CAMPGROUNDS LOCATED IN AREA 1 MAP 3

Trail No.	Trail Name	Map Loc.	Distance	Difficulty	Beginning Elev.	Ending Elev.	Ranger District
509	**Columbine Lake**	L 2	5 Mi.	Most Diff.	10,200'	12,720'	Columbine

ACCESS #1: Ophir Pass Road. Approximately 57 miles N. of Durango on Hwy. 550, turn off on 0phir Pass Road, FDR 679. Follow this down across Mineral Creek until you come to a turn off to the right. Parking is best here since parking at the immediate trailhead is limited. Follow this old mining road approximately 1 mile to the unmarked trail on the left. **ACCESS #2:** For the strong hiker and experienced route finder, there are a few other access points; Bridal Veil Basin, Ophir Pass and Porphyry Basin. **ATTRACTIONS:** This "secret" spot is usually frequented by only a few locals due to its isolated location and its difficult access. Trail consists of many steep switchbacks, then eases out into the basin below Columbine Lake. Availability of water is fairly scarce, but it is there. A few camp spots are there and tents should be moved if you plan to stay more than 2 days due to the fragile tundra. A trout dinner is a possible consideration for the fishing fan. Surrounding peaks should be climbed by experienced mountaineers only. **USE:** Light. **ACTIVITIES:** HIKING, FISHING. **USGS:** Ophir, Silverton Quads. **MAPS:** 2&3.

Trail No.	Trail Name	Map Loc.	Distance	Difficulty	Beginning Elev.	Ending Elev.	Ranger District
606	**Highland Mary**	M 2	4 Mi.	Most Diff.	8,400'	10,600'	Columbine

ACCESS #1: Follow U.S. 550 north to Silverton. Continue north through the town of Silverton toward Howardsville. Approx. 3 miles past Howardsville, turn onto FDR 589 and follow this road to the trailhead. 2WD vehicles may not be able to complete the last 3/4 mile of the drive. Parking is available in both places. The road past the wilderness boundary is closed to all motorized vehicles and mountain bikes. **ACCESS #2:** The Highland Mary Trail can also be accessed via the Whitehead Peak Trail and the Continental Divide Trail. **ATTRACTIONS:** The Highland Mary area is a beautiful high altitude subalpine basin that is for the most part, free of trees. The open meadows are filled with wildflowers in the late summer, providing beautiful views and excellent opportunities for photography. The area is used by large numbers of people and is not the place to go for solitude. The lakes provide fishing, but like most subalpine lakes, tend to remain frozen until summer. Water availability is not a problem, but all water used in this area should be filtered do to Giardia. NOT RECOMMENDED FOR HORSE USE! **USE:** Heavy. **ACTIVITIES:** HIKING, FISHING. **USGS:** Howardsville Quad. **MAP:** 3.

Trail No.	Trail Name	Map Loc.	Distance	Difficulty	Beginning Elev.	Ending Elev.	Ranger District
674	**Whitehead**	M 2	7 Mi.	Most Diff.	12,800'	11,600'	Columbine

ACCESS #1: Access the lower end of this trail via Kendall Mtn. Road from Silverton. Follow U.S. 550 from Durango to Silverton and turn off U.S. 550 onto Green St. in Silverton. Follow Green to 14th St. Turn onto 14th and follow the road eastward across the bridge. This road is the Kendall Mountain Road. About 2-1/2 miles up the road, it splits with one road heading to Kendall Peak and one going to Deer Park. Take the Deer Park Rd. (the right fork). The trailhead on this road is located approx. one mile from the fork in the road. Limited parking is available and 4WD with clearance is recommended. **ACCESS #2:** This trail can be accessed from the Highland Mary Lakes Trail and from the Continental Divide Trail. **ATTRACTIONS:** Parts of this trail are hard to follow. Most of the trail is located above timberline, so navigation with a compass is best. The alpine meadows are full of blooming wildflowers in late July and in August. Water in area is abundant, but should be purified due to problems with Giardia. As with all high altitude areas, storms can move in rapidly bringing severe lightning and rain, snow and hail. Plan accordingly and please tread lightly. **USE:** Light. **ACTIVITIES:** HIKING. **USGS:** Silverton, Howardsville Quads. **MAP:** 3.

Trail No.	Trail Name	Map Loc.	Distance	Difficulty	Beginning Elev.	Ending Elev.	Ranger District
813	**Continental Divide**	L 2	5.2 Mi.	More Diff.	12,600'	11,600'	Columbine

ACCESS #1: Access the lower end of this trail via Kendall Mountain Road from Silverton. Follow U.S. 550 from Durango to Silverton and turn onto Green St. in Silverton. Follow Green to 14th St. Turn onto 14th St. and follow the road eastward across the bridge. This road is the Kendall Mountain Road. About 2-1/2 miles up the road, it splits with one road heading to Kendall Peak and one going to Deer Park. Take the Deer Park Rd. (the right fork). The trailhead on this road is located approx. one mile from the fork in the road. Limited parking is available and 4WD with clearance is recommended. **ACCESS #2:** This trail can be accessed from the Highland Mary Lakes Trail and the Elk Park Trail. **ATTRACTIONS:** Parts of this trail are hard to follow. Most of the trail is located above timberline, so navigation with a compass is possible. The alpine meadows are full of blooming wildflowers in late July and in August. This area provides excellent views and photographic possibilities. Water can be found off the trail in most of the tributaries, but due to Giardia it should be filtered. As with all high altitude areas, storms can move in extremely rapidly, bringing severe wind, lightning, rain, snow, or hail. Plan accordingly. **USE:** Medium. **ACTIVITIES:** HIKING. **USGS:** Silverton, Howardsville, Storm King Quads **MAPS:** 3&7.

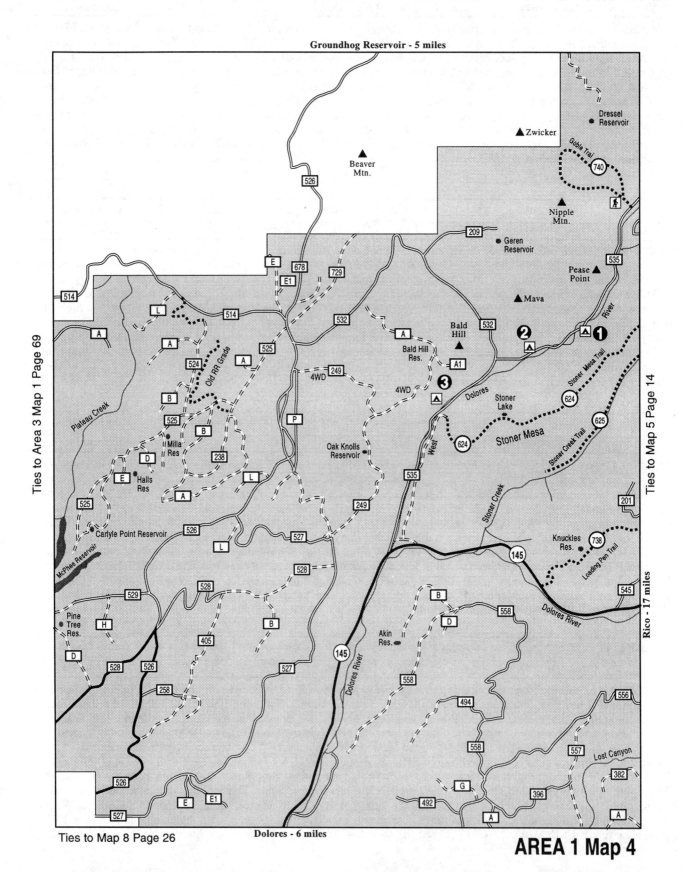

Groundhog Reservoir - 5 miles

Dressel Reservoir

▲ Zwicker

Goble Trail

(740)

▲ Beaver Mtn.

526

▲ Nipple Mtn.

209

● Geren Reservoir

535

E

678

729

Pease Point ▲

514

L

514

532

▲ Mava

River

A

524

525

A

532

A

Bald Hill ▲

532

②

A

①

Bald Hill Res.

A1

4WD

249

③

Stoner Mesa Trail

(624)

B

4WD

A

525

P

Dolores

Stoner Lake

(625)

B

238

Stoner Mesa

Stoner Creek Trail

D

L

Oak Knolls Reservoir

West

(624)

● Milla Res

E Halls Res

A

201

525

526

527

L

535

249

Stoner Creek

Knuckles Res.

(738)

(145)

Loading Pen Trail

● Carlyle Point Reservoir

528

L

528

545

529

B

558

Dolores River

● Pine Tree Res.

H

405

B

D

558

D

528

526

Akin Res. ●

258

527

558

494

E E1

558

556

526

G

557

Lost Canyon

527

(145)

Dolores River

492

396

382

A

A

Ties to Area 3 Map 1 Page 69

Ties to Map 5 Page 14

Rico - 17 miles

Old RR Grade

Plateau Creek

McPhee Reservoir

Ties to Map 8 Page 26

Dolores - 6 miles

AREA 1 Map 4

12

Trail No.	Trail Name	Map Loc.	Distance	Difficulty	Beginning Elev.	Ending Elev.	Ranger District
738	**Loading Pen**	F 4	3 Mi.	Most Diff.	7,600'	9,200'	Mancos/Dolores

ACCESS: To locate the start of Loading Pen Trail, take State Hwy. 145 north out of Dolores. Having traveled 16 miles, be on the lookout for a trail sign to your left amongst the trees. **NARRATIVE:** The beginning of this trail is made up of decomposed red sandstone and sandstone formations can be seen to the left along the trail. A creek flanks the trail to the right. You will experience a pleasant climb through a sylvan aspen grove in the first leg of the trail. Delicate scrub maple graces the creek bed. Once you emerge onto the escarpment you will find yourself among an uncrowded forest of Juniper, ponderosa pine, and scrub oak. Your view of the Dolores Valley is unimpeded. Near the summit you are afforded a grand view of the area north of Stoner. A gigantic ponderosa pine with a lightning scar stands off to the left of the trail and unusual large clumps of bracken fern are encountered west of the trail. The trail passes through an intersecting intermediate zone featuring life forms expected to be seen at lower elevations as well as denizens of the thick aspen grove. Some wildlife likely to be met along this trail include, mule deer, elk, woodpecker, and grouse.Knick-knick, mountain mahogany and service berry border the trail. As the trail widens into an old road bed you enter and area set aside for logging aspen, a logical spot to turn back or you can continue on logging roads and come out on the lower portion of Taylor Mesa on FDR 201. This trail offer an ideal moderately strenuous day hike. **USE:** Low. **ACTIVITIES:** HIKING. **USGS:** Stoner, Wallace Ranch Quads. **MAPS:** 4&5.

CAMPGROUNDS LOCATED IN AREA 1 MAP 4.

Map No.	Name	Fee	No. of Units	Max. Length	Elev.	Toilets	Water	Ranger District
1.	West Dolores	$	13	35'	7,800'	Yes	Yes	Mancos/Dolores
2.	Mavresso	$	14	--	7,600"	Yes	Yes	Mancos/Doloress
3.	Emerson	$	7	--	7,600'	Yes	Yes	Mancos/Dolores

Map Symbol Explanation

P Parking Area — Forest Service Facility — (285) U.S. Highway — National Forest Area
Picnic Area — Fishing Area — (126) State Highway — Water
Trail Head — Snowmobile Trail — (258) County Highway — Trail
Downhill Ski Area — Ice Skating Area — (1606) Trail Number — River or Stream
Boat Launch — Cross Country Ski Area — [1105] Forest Service Road — Primary Road - Paved
Bicycle Trail — Hunting — ▲ Mountain — Improved Road - Unpav
4WD Road — Campground — Colorado Trail — Unimproved Road - 4W
Motorcycle Trail — Towns & Locales — Continental Divide Trail — Forest / Wilderness Boundary

NO CAMPGROUNDS LOCATED IN AREA 1 MAP 5

Ties to Map 1 Page 6

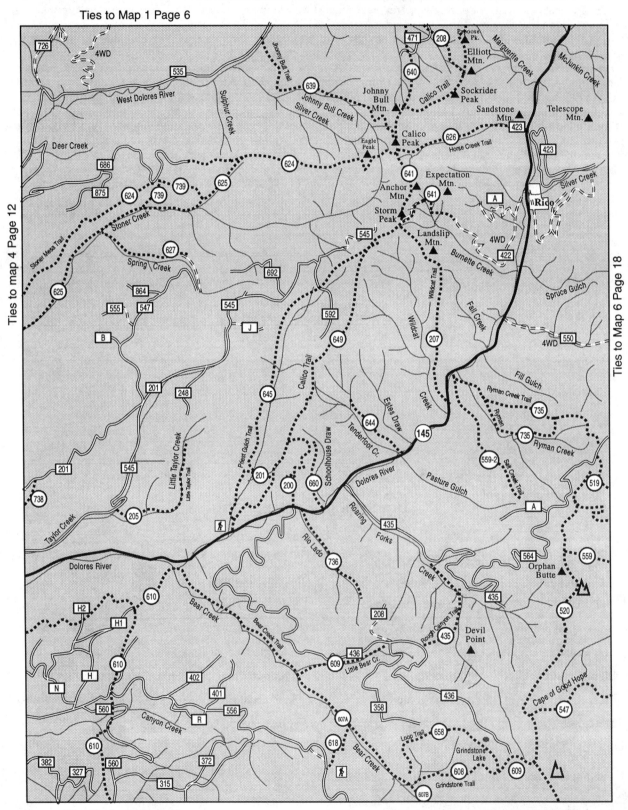

Ties to map 4 Page 12

Ties to Map 6 Page 18

Ties to Map 9 Page 26

AREA 1 Map 5

Trail No.	Trail Name	Map Loc.	Distance	Difficulty	Beginning Elev.	Ending Elev.	Ranger District
207	**Wildcat**	H 4	6.3 Mi.	Difficult	8,400'	11,800'	Mancos/Dolores

ACCESS: From Dolores, take Hwy 145 north and east. Approximately 5 miles northeast of Priest Gulch Trailhead, the highway crosses the Dolores River. The trailhead is marked by a sign on the west side before the bridge, next to the highway right-of-way fence. The road gate has been locked by the land owner so you must park next to the highway. A hiker/horse gate is located to the right of the road gate, the first 1/2 mile is on private land, please respect it. In a few yards look for the trail beginning on your left. **NARRATIVE:** This trail follows the creek bottom for about 1 1/2 miles, then switches back up to the top of the ridge to the east. Look for a trail post marking the point where it leaves the creek bottom. Cow trails are abundant and well defined in this area be alert! The trail then follows the ridge top north through a mix of aspen and spruce-fir, for about 2 miles. Then it reaches a series of meadows on the south slope of Landslip Mtn. At the head of the meadows the trail turns west, contouring around the south and west slopes of Landslip Mtn., through an old growth spruce-fir forest. After passing through a saddle just west of Landslip Mtn. the trail becomes faint, just head up along the north side of the knob to the west. The Burnett Trail lies in the saddle just to the west of it. This hike is long and steep, and water is scarce. In the later part of summer once leaving Wildcat Creek, the scenery is great on the upper 1 mile and the opportunity to see elk and deer is excellent on all the trail, especially on the upper half. The trail is primarily a foot end horseback trail, with minimal maintenance and those using it should be familiar with using Topographic maps. **USE:** Low. **ACTIVITIES:** HIKING, HORSES. **USGS:** Rico Quad, Orphan Butte. **MAP:** 5.

Trail No.	Trail Name	Map Loc.	Distance	Difficulty	Beginning Elev.	Ending Elev.	Ranger District
435	**Rough Canyon**	H 4	4 Mi.	Most Diff.	8,400'	10,400'	Mancos/Dolores

ACCESS: The trail is easily accessed either on Hillside Drive (FDR 438) and the Roaring Fork (FDR 435). The trail on Hillside Drive begins across from a corral near the end of the gravel section of FDR 436. This is approximately 12 miles from Hwy. 145. On the Roaring Fork Road look for a sign on a pull out approximately 1.3 miles up from Hwy. 145. **NARRATIVE:** Starting from Roaring Fork Road, the trail follows Roaring Fork Canyon. The trail primarily stays on the east side of the creek, but crosses to the west side for 1/2 mile at 1 mile from the start. At the mouth of Rough Canyon, the trail runs along the west side and switches back up the ridge, once on the ridge the hiker continues to climb to the top. Rough Canyon Trail truly lives up to its name. The trail is clearly defined at both upper and lower openings, the center however is far from easy going. This trail has been seldom used in recent years, so the trail is faint in spots requiring careful observation to avoid being thrown off track. Rough Canyon could be an intriguing diversion for those not discouraged by less than obvious route. An experienced hiker equipped with a topo map might enjoy traveling this beautiful and challenging trail. A variety of vegetation, aspen, spruce-fir and meadows are encountered here. Fishing is available for small Trout in Roaring Fork Creek. The trail is extremely steep once it leaves the creek, and should be avoided if possible when wet. Although suited to foot and horseback travel, motorcycle travel is discouraged due to the primitive condition of the trail. **USE:** Very Low. **ACTIVITIES:** HIKING, HORSES, FISHING. **USGS:** Orphan Butte Quad. **MAP:** 5.

Trail No.	Trail Name	Map Loc.	Distance	Difficulty	Beginning Elev.	Ending Elev.	Ranger District
607A	**Bear Creek**	G 4	13 Mi.	Easy/Mod.	7,900'	9,200'	Mancos/Dolores

ACCESS: Travel 22 miles east of Dolores, CO. on State Highway 145. Look for a green highway sign "National Forest Access Bear Creek Trail" on the right. Look for a wooden bridge crossing the Dolores River about 200 feet off the highway. You may cross the wooden bridge and park in the small parking area provided. This is private land, please stay on the road or trail, no camping or fires. The Bear Creek Trail begins in front of the corral and then goes left; the Morrison Trail is right. The Bear Creek Trail may also be accessed from the Sharkstooth Trail, Little Bear Trail, and from the Gold Run Trail and trailhead. **ATTRACTIONS:** Bear Creek is one of the most isolated and beautiful areas in the Mancos/dolores Ranger District. Life zones range from oakbrush at the mouth of the creek to aspen, spruce and fir where the trail becomes the Highline Loop National Recreation Trail. Another five miles upstream takes the visitor to the alpine zone and the headwater of Bear Creek. **NARRATIVE:** This trail parallels Bear Creek for its entire length. Hiking, mountain biking, horsebacking and trail biking are the only methods of transportation which can access Bear Creek. The first 1.5 miles of the trail is on the canyon side away from good camping spots. Fishing is allowed in Bear Creek although is limited to fly and lure only by State regulation. In 1989, the Forest Service conducted a major fish habitat improvement project in the upper 3 miles of the creek. There are no developed water supplies along the trail. Be prepared to boil or otherwise treat your drinking water. Please use good outdoor ethic in all matters of sanitation, fire and other camping skills. This trail passes through private land where livestock may be grazing. Cattle grazing is NOT allowed on National Forest land in Bear Creek, so please keep the gates closed. Grazing was halted in the mid-1980's to protect the important riparian habitat. This trail is available for use by hiker, horseback rider, mountain biker and trail bike rider. This is not a beginner's trail due to the steep pitches and loose rock. Bikers will do quite a bit of "hiking". **USE:** Light. **ACTIVITIES:** HIKING, HORSES, MOTORCYCLES, FISHING. **USGS:** Wallace Ranch, Orphan Butte Quads. **MAP:** 5.

Trail No.	Trail Name	Map Loc.	Distance	Difficulty	Beginning Elev.	Ending Elev.	Ranger District
618	**Gold Run**	G 4	2.5 Mi.	More Diff.	10,600'	9,00'	Mancos/Dolores

ACCESS: Travel 41 miles north from Mancos, CO on State Highway 184, turn east on FDR 561 for about 19 miles to the Gold Run Trailhead facilities. Trail begins at the southeast end of the facilities (near corrals). The trailhead has a restroom, hitching racks, gravel parking area and picnic tables. **NARRATIVE:** This trail primarily serves as access to Bear Creek. It meets Bear Creek Trail (FDT 607) at the bottom of Bear Creek. At this intersection the hiker may continue downstream to the mouth of Bear Creek or upstream to the Highline Loop National Recreation Trail or Sharkstooth Trail (FDT 622). The Gold Run Trail goes down a tributary side canyon into Bear Creek. It passes through stands of aspen and spruce. An open park provides vistas of a small portion of the Bear Creek Drainage. A large rockslide, about midway down the trail, can occasionally block the trail to horse traffic. This trail may be used by hikers, horseback riders or people on trail bikes. **USE:** Medium. **ACTIVITIES:** HIKING, BIKING, HORSES, MOTORCYCLES. **USGS:** Orphan Butte, Wallace Ranch Quads. **MAP:** 5.

Trail No.	Trail Name	Map Loc.	Distance	Difficulty	Beginning Elev.	Ending Elev.	Ranger District
626	**Horse Creek**	H 3	3 Mi.	Most Diff.	9,520'	11,500'	Mancos/Dolores

ACCESS: Coming from Dolores continue through Rico almost 2 miles. You will note a gate which is used to close the road in bad weather. Turn left (west) onto FDR 423 just south of this gate. This jeep road can be driven about 1 mile with high clearance vehicle before Horse Creek Trail takes off. **NARRATIVE:** This trail provides access to the Calico Trail. The intrepid hiker is rewarded with views of the La Plata and Rico ranges, Lizard Head, Eagle Peak Dunn, Middle and Dolores Peaks (to name a few). The route is noted for the dramatic climb up through thickly growing meadows, forests of aspen, spruce and fir. A lucky individual is likely to encounter elk along the fringes of timberline. Mule deer and coyote may be sighted as well as the occasional high flying eagle. The trail climbs quickly 2,000 feet in three miles. This fact accounts for the popularity of this trail with horseback riders. The rapid increase in elevation makes for a strenuous hike. In the area above the meadowlands there may be some confusion as to which trail to follow it is not clear which of two paths running parallel represents the correct route. However both of these trails continue in the same general direction and will lead back down to familiar country on the Horse Creek Trail. Approximately 1.5 miles from the trails beginning as a old mine with a vertical shaft about 10 feet off the trail, although this shaft has been closed. BE CAREFUL! **USE:** Low. **ACTIVITIES:** HIKING, HORSES. **USGS:** Rico Quad. **MAP:** 5.

Trail No.	Trail Name	Map Loc.	Distance	Difficulty	Beginning Elev.	Ending Elev.	Ranger District
639	**Johnny Bull**	G 3	4.9 Mi.	Moderate	8,450'	11,480'	Mancos/Dolores.

ACCESS: Approximately 4 miles below Dunton or 18 miles from the junction wuith Hwy 145, look for a dirt road on your right which leads down to the West Dolores River and the trailhead. The Trail begins at a sign on the West Dolores River. Note: The sign has been stolen repeatedly and may not be there. **NARRATIVE:** It will be necessary to cross the West Dolores River. The trail then passes through a stand of willows, crosses a meadow and climbs a steep hillside before following the Johnny Bull Creek drainage to the Calico Trail. The first mile is steep and primitive in places but the rest is a pleasent ascent through spruce fir forests and groves of aspen. This is the trail of choice for wildlife sightings. Our trail crew in the course of a day have seen 3 bears (a mother with cubs), a group of three majestic bull elk, and numerous deer. 28 species of birds including eagle nest within this ecosystem. After intersecting the Calico Trail scenic views of the various San Juan peaks and mesa tops can be seen by going 1/2 mile south to Calico Peak of north toward Johnny Bull Mountain. This trails pleasant ascent encourages its use by horse back as well as the hiker. **USE:** Low/ Moderate. **ACTIVITIES:** HIKING, HORSES. **USGS:** Rico, Clyde Lake Quads. **MAP:** 5.

Trail No.	Trail Name	Map Loc.	Distance	Difficulty	Beginning Elev.	Ending Elev.	Ranger District
644	**Tenderfoot**	H 4	4 Mi.	Most Diff.	8,200'	10,600'	Mancos/Dolores

ACCESS: From Dolores proceed north on Hwy. 145, continue past Priest Gulch approximately 4 miles. There is a sign next to the highway marking the trail. The trail takes off north from a right of way through private property. Be sure to securely close all the gates you pass through and stay on the trail.. **NARRATIVE:** One of the many available paths leading onto Calico, this is a fairly gradual climb except the last 1/2 mile. Losing sight of the trail is a real possibility at the final stretch. Your greatest chance of error occurs when you make your way around some grassy hills past where the ground has fallen away. Once around these hills, if you head into the timber, you've made a mistake. Though you can gradually climb up through this thickly timbered area to Calico, the correct route is up over the hills (north) there you discover blazed trees and a visible footpath. There are some quite nice views of surrounding mountains before you reach the Calico intersection. Once past the first leg of the trail where you encounter occasional steep climbing and traverse a marshy area, you will find yourself on the edge of a hillside with fine scenery to the south. Note: This is a primitive trail which may be hard to find and follow. **USE:** Low. **ACTIVITIES:** HIKING, FISHING. **USGS:** Orphan Butte, Rico Quads. **MAP:** 5.

Trail No.	Trail Name	Map Loc.	Distance	Difficulty	Beginning Elev.	Ending Elev.	Ranger District
645	**Priest Gulch**	G 4	7.5 Mi.	Difficult	8,100'	11,320'	Mancos/Dolores

ACCESS #1: 25 miles north of Dolores on on Colo. Hwy. 145 (12 miles south of Rico). Look for Priest Gulch Trailhead sign on west side of road across from Priest Gulch Store. Turn left and go .25 mi. past private property. Stay on road and do not trespass. Trailhead has parking, signing and horse facilities. **ACCESS #2:** North on Colo. Hwy. 145 19 miles to Taylor Mesa Road (FDR 545). Turn northeast and go 15 miles (Past intersection with FDR 547) to junction with FDR 592. Turn right and go 1 mile to trailhead on right. **NARRATIVE:** Trail follows Priest Gulch drainage through lush meadows and forests. Watch for a landslide 3 miles in. Trail crosses Shoas Peak Road (FDR 592) about 6 miles in. Trail then climbs steeply 1 mile to intersect with Calico Trail. Area is secluded. Deer and elk abound. The bottom of this trail follows Priest Creek and is a gentle trail suitable for those looking for an easy hike.**USE:** Light.. **ACTIVITIES:** HIKING, HORSES. **USGS:** Clyde Lake, Rico, Wallace Ranch Quads. **MAP:** 5.

Trail No.	Trail Name	Map Loc.	Distance	Difficulty	Beginning Elev.	Ending Elev.	Ranger District
649	**Calico**	G 4	13 Mi.	More Diff.	8,000'	11,720'	Manco/Dolores

ACCESS: The primary access is located near the south end of Priest Gulch Trailhead, located off Hwy 145, 12 miles south of Rico, 10 miles north of Stoner. (A portion of this trail is part of the Calico NRT, see page 7 for access and trail description.) Other trails along Hwy 145 also provide access to various points along the trail, Section House, Tenderfoot, Burnett, Johnny Bull and Horse Creek among these. The trailhead at Priest Gulch offers paved parking, signing, horse loading ramp and hitching rack. **NARRATIVE:** This trail was cut out in the 1930s by the Civilian Conservation Corp. Up to 100 men worked on widening the trail so it could be used to drive stock into the high country. Today this trail receives light use yet provides some of the best scenic views of the surrounding mountains. In the Engelmann spruce environment, associated plants include aspen, Douglas fir, common juniper, sedges, and buffaloberry. There are over 28 species of birds, including the golden eagle, that nest in this ecosystem and over 24 species of mammals from black bear and elk to the vagrant shrew, that may be found here. Above 11,000 feet the alpine ecosystem, is basically devoid of trees, instead supporting sedges, willows, fescues and numerous forbs. Permanent animal residents include the pika, yellow-bellied marmot, weasels and white-tailed ptarmigan. Mule deer and elk utilize this habitat during the summer. Migrant birds include the water pipit, rock wren, lson's warbler, and white crowned sparrow. Harsh weather conditions can and have created mudslides that require care and occasional detouring in crossing. **USE:** Low. **ACTIVITIES:** HIKING. **USGS:** Rico, Clyde Lake, Wallace Ranch Quads. **MAP:** 5.

Trail No.	Trail Name	Map Loc.	Distance	Difficulty	Beginning Elev.	Ending Elev.	Ranger District
735	**Ryman Creek**	H 3	6.8 Mi.	Most Diff.	8,600'	10,800'	Mancos/Dolores

ACCESS #1: Off of Hwy 145 near the Dolores/Montezuma County Line, look for a rough gravel road to the east. Go up the road, through the gate and park. Cross Ryman Creek and pick up the trail. **ACCESS #2:** Go east from Dolores to FDR 435. Turn right and follow it until it becomes FDR 564. Follow it north about 5 miles to the trailhead on the left side of the road. This point is a short closed logging road off FDR 564. **NARRATIVE:** This trail can be broken into two parts. Portions of the trail follows Ryman Creek (bring your fishing rod for small trout). The other part follows a ridgeline up through aspen/spruce stands to FDR 564. The part following Ryman Creek eventually fades and it is difficult to find the portion of the trail that forms a loop. From Hwy 145, the trail follows part of an old roadbed up the bottom of Ryman Creek for about 2 miles. This trail is used by hikers, horses, and due to its primitive non use experienced motorcycles are not recommended. Parts of it are very steep and rocky, especially where it climbs up steep pitches on the ridgeline portion of the trail. This area provides good habitat for mule deer and elk. It offers solitude due to its low use and roadless characteristics. This trail recieves minimum maintainance and may be hard to find and follow. **USE:** Low. **ACTIVITIES:** HIKING, HORSES, FISHING. **USGS:** Orphan Butte, Rico Quads. **MAP:** 5

Trail No.	Trail Name	Map Loc.	Distance	Difficulty	Beginning Elev.	Ending Elev.	Ranger District
739	**Twin Spring**	G 3	2.1 Mi.	Moderate	9,900'	9,800'	Mancos/Dolores

ACCESS: To reach the trailhead at Twin Spring, from Dolores take Hwy145 north approximately 13 miles. To the West Dolores Road (FDR 535) north approximately 10 miles to the Stoner Mesa Road (FDR 686), take FDR 686 east approximately 7 miles to Twin Springs. A sign marks the location. (Note: This point is near a corral and past the sharp corner from which the lower Stoner Mesa Trail begins). Park next to the road and walk down the hill, at Twin Springs Reservoir the trail forks. To the right you can go directly to Stoner Creek and the Stoner Creek Trail. To the left you can take a route to the upper reaches of the Stoner Creek trail. **NARRATIVE:** This trail passes through a rich forest of aspen and spruce/fir and occasional open meadows. It provides users a choice of access points into the Stoner Creek drainage. This area of forest is excellent summer range for deer and elk. After entering the meadows, some confusion as to location of the trail could arise. Watch for meadow post markers and cairn at these points. This trail provides the easiest way to reach the stream fishing in Stoner Creek. **USE:** Moderate. **ACTIVITIES:** HIKING. **USGS:** Clyde Lake Quad. **MAP:** 5.

CAMPGROUND INFORMATION ON PAGE 23

Ties to Map 2 Page 8

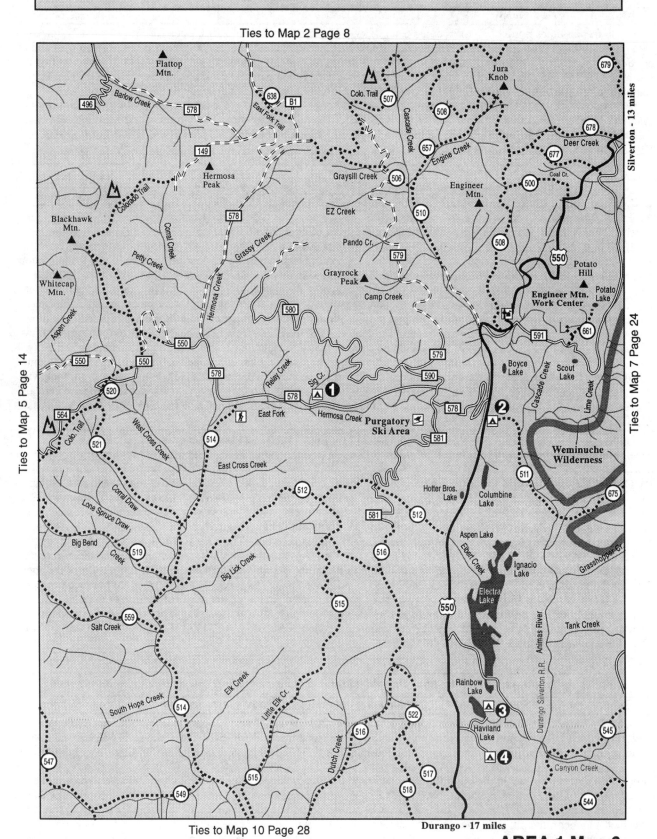

Ties to Map 5 Page 14

Ties to Map 7 Page 24

Silverton - 13 miles

Ties to Map 10 Page 28

Durango - 17 miles

AREA 1 Map 6

Trail No.	Trail Name	Map Loc.	Distance	Difficulty	Beginning Elev.	Ending Elev.	Ranger District
500	**Pass**	K 3	2.2 Mi.	More Diff.	10,600'	12,000'	Columbine

ACCESS #1: Drive approximately 30 miles north of Durango on U.S. 550 just past the pull-off area on top of Coal Bank Pass. To reach the parking area turn at "Pass Trail, Engineer Mtn." trailhead sign. **ACCESS #2:** Approximately 30 miles north of Durango on U.S. 550, turn left at Engineer Mountain trailhead. Follow this trail five miles to intersect with Pass Trail at the base of Engineer Mountain. **ATTRACTIONS:** Pass Trail provides the quickest access to Engineer Mtn. It climbs above the highway up moderate switchbacks. From there, it's a gradual climb up to the base of Engineer Mtn. Much of the trail falls on the heavily timbered northern aspect of the mountain where patches of snow often linger over the trail until mid-summer. Toward the top, the trail breaks out of the timber into some beautiful alpine meadows at the base of Engineer Mtn. If you don't want to stop hiking, follow the Engineer Mtn. Trail north towards Jura Knob. A nice loop trip could be planned with Coal Creek or Deer Creek Trails. Water availability is good, but should be filtered to avoid problems with Giardia **USE:** Heavy. **ACTIVITIES:** HIKING. **USGS:** Engineer Mountain Quad. **MAP:** 6.

Trail No.	Trail Name	Map Loc.	Distance	Difficulty	Beginning Elev.	Ending Elev.	Ranger District
506	**Graysill**	K 3	3.0 Mi.	More Diff.	9,600'	11,720'	Columbine

ACCESS: Access either end of the trail by Cascade Divide Road via Purgatory or Cascade Creek Trail. **ATTRACTIONS:** This trail provides a quick and easy way out of Cascade Canyon. From the bottom of Cascade Creek, the trail backs up a north facing slope to Cascade Divide Road. Trail is sign posted through old timber cuts below the road. There is no water on this trail. The main attraction from this trail is to get to the Rico-Silverton Trail instead of going all the way to the end of Cascade Creek. **USE:** Light. **ACTIVITIES:** HIKING. **USGS:** Engineer Mountain Quad. **MAP:** 6.

Trail No.	Trail Name	Map Loc.	Distance	Difficulty	Beginning Elev.	Ending Elev.	Ranger District
508	**Engineer Mountain**	K 3	11 Mi.	More Diff.	9,000'	12,000'	Columbine

ACCESS #1: 31 miles north of Durango on U.S. 550 from Engineer Mtn. Guard Station. To reach station, turn west on FDR 817 2/3's of a mile past Cascade Village and follow this road 100 yds. The trailhead is on the north side of the road just past the cattle guard. There is adequate parking along the dirt road. **ACCESS #2:** Via Pass Creek Trail (FDT 500). Approx. 37 miles north of Durango on U.S. 550 at the top of Coal Bank Pass turn west on the dirt road and follow it 50 yds. to the trailhead. Follow this trail 2 miles to the junction with the Engineer Mountain Trail. **ATTRACTIONS:** Water availability is limited on this trail. The trail is in good shape and maintains an even uphill grade. This trail offers access to Engineer Mtn. and The possibility for loop hikes to get views of both the lower valley and the Needle and Grenadier Mountain Ranges. The trail is located primarily on south facing slopes and is usually accessible during late spring. Both of the trailhead access points are good slopes for seeing native wildflowers in July and August. The trail begins in a spruce-fir forest and gradually winds up to subalpine conditions. **USE:** Medium. **ACTIVITIES:** HIKING. **USGS:** Engineer Mountain Quad. **MAP:** 6.

Trail No.	Trail Name	Map Loc.	Distance	Difficulty	Beginning Elev.	Ending Elev.	Ranger District
510	**Cascade Creek**	K 3	4.7 Mi.	More Diff.	9,200'	9,720'	Columbine

ACCESS: 30 miles N. of Durango on U.S. 550, turn west on the Cascade dirt road (frontage access sign). This road eventually turns to 4WD but there is a little cul-de-sac area just past the flume where there is some parking. From here follow the dirt road approximately 3/4 mile to a red gate before a cabin on the right. Open the gate and follow the road until it turns to the trail. **ATTRACTIONS:** There are two trails on either side of the creek. The trail on the east is a stock driveway that is steep and muddy in spots. The trail on the west is more pleasant, it stays closer to the stream and is moderate hiking. Where Engine Creek Trail veers north away from the trail. Directly across the creek is where the westside trail ends (or turns into Graysill Trail). It is best to cross the creek up a little ways. Trail deadends with Graysill and Engine Creek Trails. There are several beautiful views and waterfalls. Plenty of water is available, but purification is recommend to avoid Giardia. **USE:** Medium. **ACTIVITIES:** HIKING. **USGS:** Engineer Mountain Quad. **MAP:** 6.

Trail No.	Trail Name	Map Loc.	Distance	Difficulty	Beginning Elev.	Ending Elev.	Ranger District
512	**Elbert Creek**	J 4	9.9 Mi.	More Diff.	8,800'	8,400'	Columbine

ACCESS #1: 22 miles north of Durango on U.S. 550, turn off of the highway southwest of the Needles Country Store. The trailhead is located at the corrals. Be sure to close the gates as you go through. Parking is available. **ACCESS #2:** Turn off onto FDR 578, which is located 29 miles north of Durango on U.S. 550 and follow 2-1/2 miles to a fork. Take a left turn onto FDR 581 and follow it for about 3-1/2 miles to Elbert Creek and the trailhead. **ACCESS #3:** Access is available via the Hermosa Trail. Follow FDR 578 for approx. 9 miles to the Hermosa Creek Trailhead in Hermosa Park. Follow the trail approx. 5 miles downstream to the Big Lick Creek and Elbert Creek Trail junction. **NARRATIVE:** This trail provides access into the Hermosa drainage. Day trips can be planned from access points 1 and 2 and can be utilized to climb Castle Rock. From the Needles Country Store, the trail climbs steadily uphill on an east facing slope inhabited by aspens. A spring surfaces near the cabin at the base of Castle Rock. From there, the trail curves around the base of Castle Rock and continues to climb. The trail then crosses FDR 581 and climbs a small hill, traversing beautiful, scenic meadows. The trail then abruptly drops for approx. 7 miles into the Hermosa drainage. Water is available, but requires filtering to prevent possible Giardia. **USE:** Medium. **ACTIVITIES:** HIKING. **USGS:** Electra Lake, Elk Creek Quads. **MAP:** 6.

Trail No.	Trail Name	Map Loc.	Distance	Difficulty	Beginning Elev.	Ending Elev.	Ranger District
515	**Little Elk Creek**	J 4	13 Mi.	More Diff.	10,400'	7,800'	Columbine

ACCESS: Little Elk Trail can be accessed by taking U.S. 550 from Durango, 26 miles north to Purgatory Ski Area. From the ski area take Hermosa Park Road (FDR 578) to Elbert Creek Road (FDR 581) turn left (south) and follow the main dirt road approximately 8 miles. The road ends in a cul-de-sac (plenty of parking) and the trailhead is in the south corner of the cul-de-sac. **ATTRACTIONS:** This is a quiet trail that offers beautiful views. Walking the first few miles, the La Plata Mountains, Hermosa Valley and Animas Valley are within view. There is no water during the first few miles of the trail, but when you reach the cabin there is a nice creek a few yards to the east. As with most water in the forest, it should be filtered to avoid Giardia. The trail then follows this creek for a few miles before veering towards the east to tie in with the Hermosa Trail. **USE:** Medium. **ACTIVITIES:** HIKING. **USGS:** Elk Creek, Electra Lake, Monument Hill Quads. **MAPS:** 6&10.

Trail No.	Trail Name	Map Loc.	Distance	Difficulty	Beginning Elev.	Ending Elev.	Ranger District
516	**Dutch Creek**	J 4	9.7 Mi.	More Diff.	10,280'	8,000'	Columbine

ACCESS #1: Elbert Creek Road (FDR 581) above-Purgatory, access from Hermosa Park Road (FDR 578), off of U.S. 550 26 miles north of Durango, just north of Purgatory entrance. Follow FDR 578 to first major turn-off and take a left FDR 581. Follow FDR 581 just over 8 miles to trailhead. **ACCESS #2:** Take Hwy. 550 N. 10 miles to Hermosa and turn W. on C.R. 201 and keep right. Road will become FDR 576. Follow this road for approximately 3 miles. Trailhead on the northside of road with adequate parking. **NARRATIVE:** From access #1, Dutch Creek Trail starts high and ends low. For the first few miles it basically follows a ridgeline at 10,000 ft. Within 3.5 miles the trail tops out in open meadows with wide ranging views to both the east and west. To the west one can see out over the width of the Hermosa country for a clear view of the La Platas. To the east one can see the Needle Mountains. Coming down off the ridge, the trail becomes obscure in places dropping down into Stag Draw. Within two miles, the tail intercepts Dutch Creek for an accessible water supply. Grassy hillsides and a few fairly open parks along Dutch Creek provide the best campsites. Once the trail breaks off at Dutch Creek, access to either water or grazing becomes limited. **USE:** Heavy. **ACTIVITIES:** HIKING. **USGS:** Electra Lake, Elk Creek, Hermosa, Monument Hill Quads. **MAPS:** 6&10.

Trail No.	Trail Name	Map Loc.	Distance	Difficulty	Beginning Elev.	Ending Elev.	Ranger District
517	**Goulding Creek**	K 4	2.7 Mi.	More Diff.	7,840'	10,080'	Columbine

ACCESS: Approx. 17 miles north of Durango on U.S. 550, 1/4 mile past the Hermosa Cliff Fire Station there is a dirt road and a sign marking the head of the trail. Follow the dirt road south for 100 yards and park next to the old, abandoned car. There should be 4-5 spaces available. **NARRATIVE:** This trail provides access to the Pinkerton-Flagstaff Trail and also to the eastern half of Hermosa Creek. The trail traverses up several steep switchbacks for the first mile and a half and then levels out. It then travels through some open meadows that are full of native plants and flowers and continues slightly uphill through aspen groves to a cabin near the head of the Pinkerton-Flagstaff Trail. Water is available in several streams, but should be filtered due to problems with Giardia in the area. **USE:** Medium. **ACTIVITIES:** HIKING. **USGS:** Electra Lake Quad. **MAP:** 6.

Trail No.	Trail Name	Map Loc.	Distance	Difficulty	Beginning Elev.	Ending Elev.	Ranger District
519	**Big Bend**	J 4	6 Mi.	More Diff.	10,800'	8,400'	Columbine

ACCESS: #1: 28 miles north of Durango on U.S. 550 to Purgatory Ski area, turn west to Hermosa Park Road (FDR 578). Follow for 10.5 miles to junction with Hotel Draw Road. Turn west and follow for approximately 5 miles to Scotch Creek Road junction. Directly after the junction is a gate with Highline Trail (Colorado Trail) leading from it. The trailhead is approximately 3 miles south of FDR 564 from the junction of Scotch Creek Road and FDR 564. **ACCESS #2:** Access bottom of trail from Hermosa Creek Trail (FDR 514). To reach trailhead for Big Bend, hike approximately 4 miles on Hermosa Creek Trail and cross over the bridge. **ATTRACTIONS:** There are wonderful fishing holes, beautiful views and several types of wildflowers. The entire drainage is usually full of elk, coyote and an occasional bear. Water should always be purified to avoid problems with Giardia. **NARRATIVE:** This trail would make an excellent loop trip as would most other trails in the Hermosa drainage. The trail follows a stream except for the top part which follows a ridge. **USE:** Light. **ACTIVITIES:** HIKING, FISHING. **USGS:** Orphan Butte, Elk Creek Quads **MAPS:** 5&6.

Trail No.	Trail Name	Map Loc.	Distance	Difficulty	Beginning Elev.	Ending Elev.	Ranger District
521	**Corral Draw**	J 4	5 Mi.	More Diff.	10,880'	8,555'	Columbine

ACCESS #1: (Road access) 28 miles N. of Durango on U.S. 550, turn west at Purgatory Ski area. From there take Hermosa Park Rd. 10.5 miles to a junction with Hotel Draw Rd. Turn west on this road and follow it approximately 5 miles to junction with Scotch Creek Rd. Directly after this "Y" in the road, the Colorado Trail is south from a gate. There is adequate parking for several vehicles. Follow the Colorado Trail 1 mile to the Corral Draw trailhead. 4WD recommended. **ACCESS #2:** (Trail access) Access the bottom of this trail from Hermosa Park RD. (FDR 578). The trailhead is very accommodating with bathrooms, parking, livestock ramps and an interpretive map. Follow FDT 514 for about 2.5 miles to the junction of FDT 521. **NARRATIVE:** Starting from the Colorado Trail (Highline) the trail switchbacks down and east facing grassy bowl, dipping into several gullies and soon merging into the spruce and fir at the bottom of the draw. The trail stays in the bottom of the draw for the duration of the hike. Near the bottom, the trail passes through several small south facing meadows that offer nice camping spots. Wading Hermosa Creek is necessary to access Hermosa Trail. Water should always be purified to avoid Giardia. **USE:** Light. **ACTIVITIES:** HIKING. **USGS:** Elk Creek, Hermosa Peak Quads. **MAP:**

Trail No.	Trail Name	Map Loc.	Distance	Difficulty	Beginning Elev.	Ending Elev.	Ranger District
559	**Salt Creek**	J 4	6 Mi.	More Diff.	10,800'	8,200'	Columbine

ACCESS #1: Go 28 miles north of Durango on U.S. 550 to Purgatory Ski Area, turn west and take Hermosa Park Rd. 10.5 miles to junction with Hotel Draw Rd. Turn west on this road and follow it approximately 5 miles to the junction with Scotch Creek Rd. Directly after this "Y" is a gate with the Colorado Trail leading south. There is adequate parking for several vehicles and 4WD is recommended. Follow the Colorado Trail approximately 6 miles to the Salt Creek Trailhead. **ACCESS #2:** Access the bottom of this trail from the Hermosa Creek Trail. To reach the trailhead for Salt Creek, the hiker must hike about 5 miles down the upper Hermosa Creek Trail. **ACCESS #3:** Salt Creek Trail can also be accessed by parking at the second road closure gate off Road 564 near Orphan Butte. Walk behind the gate and on the closed road to the ridge line beyond and the Colorado Trail. Look for the Salt Creek Trail sign along the Colorado Trail. **NARRATIVE:** By starting from the Highline, the hiker drops over 2,000 feet and enjoys an all downhill excursion. The top of the trail was reconstructed in 1989 and so the tread is in excellent condition. It switchbacks down through meadows and aspen glades. The bottom of the trail is a bit rougher and steeper, so watch your step. The trail goes along the creek the last few miles and offers views that seem almost tropical. At the bottom of the trail, there is an old log crossing that could be dangerous during Spring runoff. After crossing the creek, the trail merges with the Hermosa Creek Trail. **USE:** Light. **ACTIVITIES:** HIKING. **USGS:** Elk Creek, Orphan Butte Quads. **MAPS:** 5&6.

Trail No.	Trail Name	Map Loc.	Distance	Difficulty	Beginning Elev.	Ending Elev.	Ranger District
657	**Engine Creek**	K 3	7 Mi.	Most Diff.	12,200'	9,600'	Columbine

ACCESS #1: U.S. 550 north of Silverton approximately 2 miles. Take a left on FDR 585 to South Mineral Creek. Go about 2 miles past South Mineral Campground and turn left at fork in road. Rico-Silverton Trail begins at end of road. Engine Creek Trail breaks via FDT 506 off to the east about 3 miles up the Rico-Silverton in an alpine saddle. **ACCESS #2:** Approximately 30 miles north of Durango on U.S. 550, go to hairpin turn just past Cascade Village. Turn left (west) on Cascade Creek and go to the end of the summer homes. Take Cascade Creek Trail about 4 miles. Engine Creek Trail sign marks turn-off to the right. **ATTRACTIONS:** Getting to the trail from either end involves some preliminary hiking. Therefore, an overnight stay may be helpful. There is a good central location and campsite at timberline where the trail starts down along Engine Creek, along with a beautiful view of Engineer Mountain and Jura Knob. Water is abundant, but purification is recommended to avoid problems with Giardia. **USE:** Light. **ACTIVITIES:** HIKING. **USGS:** Engineer

Trail No.	Trail Name	Map Loc.	Distance	Difficulty	Beginning Elev.	Ending Elev.	Ranger District
661	**Spud Lake**	K 3	1 Mi.	Easy	9,400'	9,800'	Columbine

ACCESS: On Hwy. 550, drive approximately 28 miles north of Durango and turn east on Old Lime Creek Road (FDR 591). Follow this dirt road approximately 3 1/2 miles until you come to a fairly big, marshy lake covered with lily pads. The trailhead is just around the corner on the left with plenty of parking spots. **ATTRACTIONS:** This trail is short and easy. Passing through aspen glades, it offers excellent views of Engineer Mountain, Grayrock Peak, and Spud Mountain, along with good views of the Needle Mountains to the east. The lake provides excellent fishing and a pleasant place to relax. This is a great summer hike for all ages and the aspen leaves during the fall are spectacular. **USE:** Heavy. **ACTIVITIES:** HIKING, FISHING. **USGS:** Engineer Mountain Quad. **MAP** 6.

Trail No.	Trail Name	Map Loc.	Distance	Difficulty	Beginning Elev.	Ending Elev.	Ranger District
675	**Animas River**	K 4	6.7 Mi.	More Diff.	7,700'	8,300'	Columbine

ACCESS #1: This trail is a section of trail linking Purgatory Trail and Needle Creek Trail together. The first trail is the Purgatory Trail which begins at the Purgatory campground located 28 miles north of Durango on U.S. 550, across the road from the ski area. **ACCESS #2:** This trail can also be accessed from the Needle Creek Trail. **ATTRACTIONS:** The Animas River Trail is an alternative to taking the train to get to Chicago Basin. The trail follows the river banks for its entire distance offering fishing possibilities. The river canyon walls are steep and rocky, making the sun come up late and set early. There are many good campsites along the way. Since this is a major drainage, all water used should be filtered to prevent possible Giardia contamination. This trail is used by horse outfitters, so expect to see horse traffic. **USE:** Heavy. **ACTIVITIES:** HIKING, HORSES, FISHING. **USGS:** Electra Lake, Mountain View Crest Quads. **MAPS:** 6&7.

Trail No.	Trail Name	Map Loc.	Distance	Difficulty	Beginning Elev.	Ending Elev.	Ranger District
677	**Coal Creek**	K 3	3 Mi.	More Diff.	10,200'	11,600'	Columbine

ACCESS #1: Take U.S. 550 for approx. 37 miles north of Durango. Park in the dirt pull-out just past the first hairpin turn after Coal Bank Pass. The trail leads uphill from the northside of the road. A sign marks the Coal Creek trailhead. **ACCESS #2:** Access to this trail is also available from the Engineer Mtn. Trail. **ATTRACTIONS:** The Coal Creek Trail climbs up through fairly steep terrain, passing through a predominantly coniferous forest. There are occasional breaks in the trees that offer scenic views of the West Needles. The last section of the trail is blaze posted to make it visible as it crosses swampy terrain. The trail never comes close to Coal Creek but the Engineer Mtn. Trail crosses the creek at timberline. All water obtained in the area should be purified to prevent problems with Giardia. The Coal Creek Trail offers several possibilities for doing loop trails utilizing other trails in the same area. **USE:** Medium. **ACTIVITIES:** HIKING. **USGS:** Engineer Mountain Quad. **MAP:** 6.

Trail No.	Trail Name	Map Loc.	Distance	Difficulty	Beginning Elev.	Ending Elev.	Ranger District
678	**Deer Creek**	K 3	4.5 Mi.	Most Diff.	9,800'	11,200'	Columbine

ACCESS: Approximately 38 miles north of Durango on U.S. 550, drive into the hairpin turn and over Deer Creek, continue about an 1/8 of a mile to the dirt pull-off on the south side of the road. The trailhead is on the northside of the road.
ATTRACTIONS: Deer Creek provides some excellent views of the West Needle Mountains along with moderate hiking and good access to other trails. This trail is usually just a day hike, but can be combined with other trails such as Coal Creek and Engineer Mountain Trails for an overnighter. The trail ascends through some aspens, then switchbacks up a gully right above a small limestone cliff band. The trail continues through some nice meadows parallel to Deer Creek, then it more or less disappears. The other option would be to keep going straight up the valley/bowl to Engineer Mountain Trail. Filter all water due to possible problems with Giardia. **USE:** Light. **ACTIVITIES:** HIKING, FISHING. **USGS:** Engineer Mountain Quad. **MAP:** 6.

Trail No.	Trail Name	Map Loc.	Distance	Difficulty	Beginning Elev.	Ending Elev.	Ranger District
679	**West Lime Creek**	K 3	3 Mi.	More Diff.	9,800'	10,900'	Columbine

ACCESS #1: This access point is found approx. 39 miles north of Durango on U.S. 550. The trailhead is located approx. 1 mile north from the low point between Coal Bank and Molas Pass on the big sweeping hairpin turn. Parking is located on the eastside of the road in the large turn-out south of the trailhead. To reach the trailhead, one must cross the highway and walk approx. 40 yards up the road. The trailhead leads off on the northside of the creek. **ACCESS #2:** This trail is also accessible via the Rico-Silverton Trail, using a topo map and compass. **ATTRACTIONS:** This trail follows West Lime Creek for several miles, then crosses and leaves it. There are several waterfalls along the creek. Usually there is an abundance of wildflowers in July and August. All water from the creek that is used should be filtered to prevent problems with Giardia. The trail fades out below the Twin Sisters, but with a compass and a topographic map it is possible to navigate a route to the Rico-Silverton Trail. From here you can head north or south climbing either the Twin Sisters or another high point provides a spectacular view and excellent opportunities for photography. **USE:** Medium. **ACTIVITIES:** HIKING. **USGS:** Engineer Mountain, Ophir Quads. **MAPS:** 2&6.

Map No.	Name	Fee	No. of Units	Max. Length	Elev.	Toilets	Water	Ranger District
CAMPGROUNDS LOCATED IN AREA 1 MAP 6. (Page 18)								
1.	Sig Creek	$	8	35'	9,000'	Yes	Yes	Columbine
2.	Pugatory	$	14	35'	8,800'	Yes	Yes	Columbine
3.	Haviland Lake	$	45	45'	8,000'	Yes	Yes	Columbine
4.	Chris Park	$	Group	30-50'	8,000'	Yes	Yes	Columbine

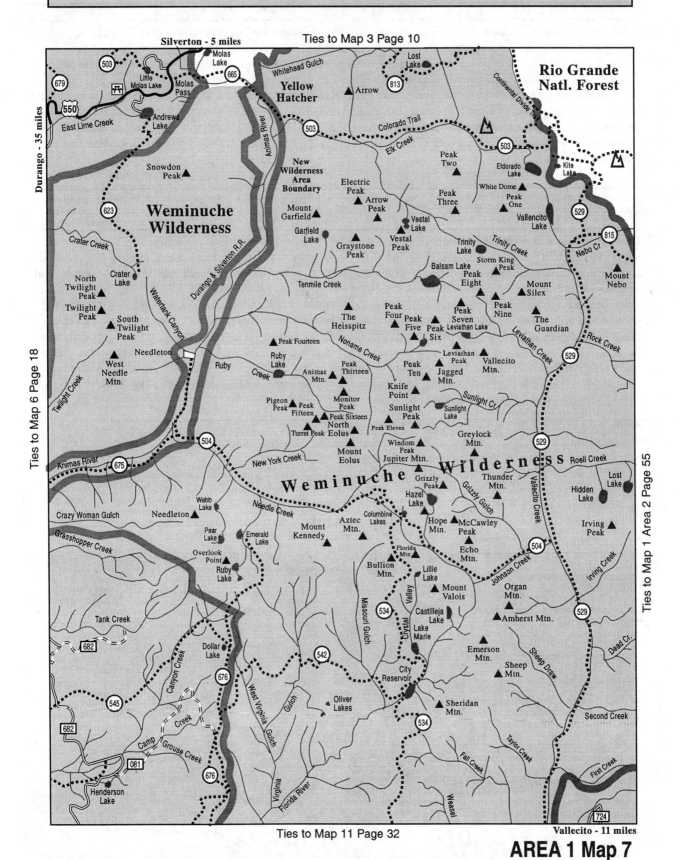

NO CAMPGROUNDS LOCATED IN AREA 1 MAP 7

Ties to Map 3 Page 10

Silverton - 5 miles

Rio Grande Natl. Forest

Durango - 35 miles

Ties to Map 6 Page 18

Ties to Map 1 Area 2 Page 55

Molas Lake
Little Molas Lake
Molas Pass
Andrews Lake
East Lime Creek
Snowdon Peak
Weminuche Wilderness
Crater Creek
North Twilight Peak
Twilight Peak
Crater Lake
South Twilight Peak
Needleton
Ruby
West Needle Mtn.
Twilight Creek
Waterlank Canyon
Durango & Silverton R.R.
Animas River
Whitehead Gulch
Yellow Hatcher
Arrow
Colorado Trail
Elk Creek
New Wilderness Area Boundary
Electric Peak
Arrow Peak
Mount Garfield
Garfield Lake
Graystone Peak
Tenmile Creek
The Heisspitz
Noname Creek
Peak Fourteen
Ruby Lake
Animas Mtn.
Peak Thirteen
Pigeon Peak
Peak Fifteen
Monitor Peak
Peak Sixteen
North Eolus
Turret Peak
Mount Eolus
New York Creek
Lost Lake
Peak Two
Peak Three
Vestal Lake
Vestal Peak
Trinity Lake
Trinity Creek
Balsam Lake
Peak Eight
Storm King Peak
Peak Four
Peak Five
Peak Six
Peak Seven
Leviathan Lake
Peak Nine
Mount Silex
Leviathan Peak
Leviathan Creek
Vallecito Mtn.
The Guardian
Peak Ten
Jagged Mtn.
Knife Point
Sunlight Peak
Sunlight Lake
Peak Eleven
Sunlight Ct.
Windom Peak
Greylock Mtn.
Jupiter Mtn.
Eldorado Lake
White Dome
Peak One
Vallencito Lake
Nebo Cr
Mount Nebo
Rock Creek
Kite Lake
Roell Creek
Lost Lake
Hidden Lake
Irving Peak
Irving Creek

Weminuche Wilderness

Webb Lake
Needleton
Crazy Woman Gulch
Grasshopper Creek
Pear Lake
Emerald Lake
Overlook Point
Ruby Lake
Needle Creek
Mount Kennedy
Aztec Mtn.
Columbine Lakes
Grizzly Peak
Hazel Lake
Hope Mtn.
McCawley Peak
Echo Mtn.
Thunder Mtn.
Grizzly Gulch
Vallecito Creek
Florida Mtn
Lillie Lake
Bullion Mtn.
Mount Valois
Organ Mtn.
Amherst Mtn.

Tank Creek
Dollar Lake
Canyon Creek
Oliver Lakes
City Reservoir
Castilleja Lake
Lake Marie
Crystal Valley
Missouri Gulch
Emerson Mtn.
Sheep Mtn.
Sheep Draw
West Virginia Gulch
Grouse Creek
Camp Creek
Henderson Lake
Virginia Gulch
Florida River
Sheridan Mtn.
Second Creek
Taylor Creek
Fall Creek
Weasel
First Creek

Ties to Map 11 Page 32

Vallecito - 11 miles

AREA 1 Map 7

Trail No.	Trail Name	Map Loc.	Distance	Difficulty	Beginning Elev.	Ending Elev.	Ranger District
503	**Elk Creek**	M 3	8.9 Mi.	Most Diff.	9,000'	12,800'	Columbine

ACCESS #1: Via Molas Trail 665 from U.S. 550, 42 miles north of Durango, off the top of Molas Pass towards Silverton. Turn east (Right) at Molas Lake. Hike 5 miles down Molas Trail to Elk Park. **ACCESS #2:** By Narrow Gauge Railroad, to "Elk Park". **ATTRACTIONS:** One of the biggest drawing points of this trail is the unique opportunity to combine a wilderness trip along with a ride on a steam locomotive. This same attraction can also effect other aspects of your wilderness experience by decreasing overall solitude with increased visitation. Elk Creek drains the northern portion of the Grenadier Range which includes Arrow, Vestal, and the Trinity Peaks, a mecca for mountaineers. These features along with access to the Continental Divide Trail make Elk Creek one of the heaviest used trails in the entire Weminuche Wilderness. Remember to filter water and practice low impact camping. **USE:** Heavy. **ACTIVITIES:** HIKING. **USGS:** Storm King, Snowdon Peak Quads. **MAP:** 7.

Trail No.	Trail Name	Map Loc.	Distance	Difficulty	Beginning Elev.	Ending Elev.	Ranger District
504	Needle Crk/Chicago Basin	L 4	8.3Mi.	More Diff.	8,000'	9,200'	Columbine

ACCESS #1: Via Purgatory Trail (FDT 675). **ACCESS #2:** Durango-Silverton Train to Needleton drop-off. **ATTRACTIONS:** Needle Creek Trail follows the creek up into Chicago Basin which is a spectacular mountain basin rimmed by three fourteen thousand foot peaks. Most of the basin is at or above timberline. This beauty, coupled with the access by the train, creates one of the heaviest used portions of the Weminuche Wilderness. Heavy use has become a major concern, so be prepared on a busy day in August to see up to 60 visitors. An off-season trip is a good idea. However, don't plan too early or late since winter can become a factor either time. If you plan to be in the Chicago Basin area long, the concept of low impact camping is of prime importance such as using stoves only, **_no fires!_** Please camp away from the water source and away from sight of the trail. Tread lightly on these fragile subalpine areas; leaving no trace. Large groups should use dispersed camping techniques to minimize impact and everyone should filter their water. **USE:** Heavy. **ACTIVITIES:** HIKING. **USGS:** Snowden Peak, Mountain View Crest, Columbine Pass Quads. **MAP:** 7.

Trail No.	Trail Name	Map Loc.	Distance	Difficulty	Beginning Elev.	Ending Elev.	Ranger District
623	**Crater Lake**	L 3	5.5 Mi.	Difficult	10,800'	11,600'	Columbine

ACCESS: Approx. 45 miles north of Durango on U.S. 550, take the Andrews Lake turn-off. This dirt road is located approx. 1/4 of a mile north of the Lime Creek rest stop which is on the eastern side of U.S. 550. The turn-off road is approx. one mile long and terminates at Andrews Lake where there is fishing and plenty of parking. Restroom facilities are also available here. The trail leads off from the southern end of the lake. **ATTRACTIONS:** The Crater Lake Trail is an excellent long day hike or an overnight trip. The lake itself is nestled in a basin at the base of Twilight Peak, making the peak readily accessible. Snowdon Peak is also accessible from the Crater Lake Trail. Climbing either of these peaks affords a spectacular panoramic view including the Needles, the Grenadiers, and the distant town of Silverton. The beginning section of the trail switchbacks up a 500 ft. rise, but once that has been ascended, the rest of the trail proceeds gently through alternating heavy timber and open meadows. Toward the end of the trail one will encounter several marshy ponds and the trail becomes somewhat indistinct, but with a keen eye, the trail is discernible. Crater Lake offers fly and spin-casting fishing. **USE:** Heavy. **ACTIVITIES:** HIKING, FISHING. **USGS:** Snowdon Peak Quad. **MAP:** 7.

Trail No.	Trail Name	Map Loc.	Distance	Difficulty	Beginning Elev.	Ending Elev.	Ranger District
665	**Molas**	L 3	4 Mi.	More Diff.	10,000'	8,900'	Columbine

ACCESS: Drive 46 miles north of Durango on U.S. 550, approx. 1 mile north of Molas Pass, turn east into the Molas Lake access road. Plenty of parking is available next to the highway and at the trailhead. The trail leads off from the southwestern side of the lake. **ATTRACTIONS:** This trail is nice to walk down, but walking up is work. Fortunately, there are other options such as the Elk Creek Trail (FDT 503) and the D.&S.N.G. train. The first part of the trail gently descends through some pretty meadows next to the creek. This section of the trail offers excellent views of the Grenadiers and Needles. The trail then begins to sharply drop down the 1,500 foot descent to the Animas River. All water used out of this area should be purified due to problems with Giardia. **USE:** Heavy. **ACTIVITIES:** HIKING. **USGS:** Snowdon Peak Quad. **MAP:** 7.

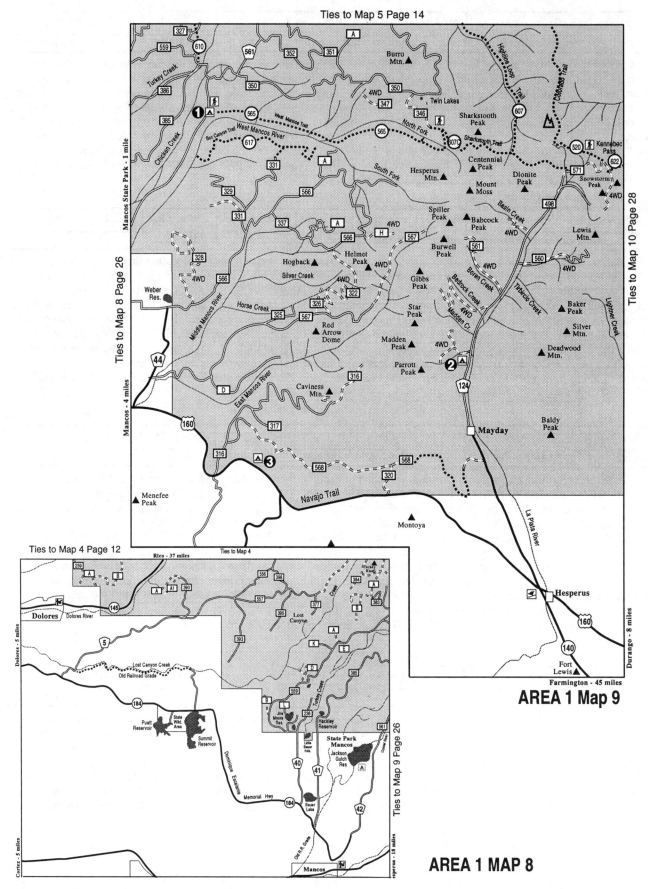

Ties to Map 5 Page 14

Ties to Map 8 Page 26

Mancos State Park - 1 mile

Ties to Map 10 Page 28

Turkey Creek

Chicken Creek

West Mancos Trail

West Mancos River

Box Canyon Trail

North Fork

4WD

Twin Lakes

Burro Mtn.

Highline Loop Trail

Colorado Trail

Sharkstooth Peak

Sharkstooth Trail

Kennebec Pass

South Fork

Hesperus Mtn.

Centennial Peak

Dionite Peak

Snowstorm Peak

4WD

Mount Moss

Spiller Peak

Babcock Peak

Basin Creek

Lewis Mtn.

Burwell Peak

Boren Creek

Helmet Peak

Hogback

Silver Creek

4WD

Gibbs Peak

Bedrock Creek

Baker Peak

Middle Mancos River

Horse Creek

Red Arrow Dome

Star Peak

Madden Cr.

Tabirdo Creek

Lightner Creek

Silver Mtn.

Weber Res.

Mancos - 4 miles

Madden Peak

4WD

Deadwood Mtn.

East Mancos River

Caviness Mtn.

Parrott Peak

Mayday

Baldy Peak

Menefee Peak

Navajo Trail

Montoya

La Plata River

Ties to Map 4 Page 12

Rico - 37 miles

Ties to Map 4

Turkey Knob

Dolores - 5 miles

Dolores

Dolores River

Lost Canyon

Lost Canyon Creek

Old Railroad Grade

Turkey Creek

Hesperus

Durango - 8 miles

Fort Lewis

Farmington - 45 miles

AREA 1 Map 9

Puett Reservoir

State Wild. Area

Summit Reservoir

Dominguz Escalante

Memorial Hwy

Joe Moore Res.

Hackley Reservoir

State Park Mancos

Jackson Gulch Res.

Ties to Map 9 Page 26

Bauer Lake

Little Bauer Res.

Mancos

Cortez - 5 miles

Hesperus - 18 miles

AREA 1 MAP 8

Trail No.	Trail Name	Map Loc.	Distance	Difficulty	Beginning Elev.	Ending Elev.	Ranger District
553	Kennebec Pass-Junction Crk.	H 5	14.5 Mi.	Moderate	11,600'	6,960'	Columbine

ACCESS: Follow U.S. 160 11 miles west from Durango to Forest Access La Plata Canyon Road, Cty. Rd. 124. Go north for 3-1/2 miles to Mayday. The gravel road can be ridden from the vicinity of Mayday, or bikes can be transported in a high clearance vehicle to Kennebec Pass, an additional 10 miles. **ATTRACTIONS:** This is a popular mountain bike loop near Durango that, starting at Mayday, makes for a difficult workout or starting at Kennebec Pass makes a long downhill run with most of the route on gravel road. **Note:** Care should be taken to make certain that you are indeed on the Sliderock trail from Kennebec Pass which heads generally east toward the Junction Creek Road. Numerous other trails in the vicinity could lead one astray. A common mistake is to head north into the South Fork of Hermosa Creek. **USE:** Heavy. **ACTIVITIES:** HIKING, MOUNTAIN BIKES. **USGS:** La Plata, Monument Hill, Durango West Quads. **MAPS:** 9&10. NOTE: Trail as described is a road route. Trail number will not appear on maps 9 or 10.

Trail No.	Trail Name	Map Loc.	Distance	Difficulty	Beginning Elev.	Ending Elev.	Ranger District
565	**West Mancos**	G 5	10 Mi.	Moderate	8,500'	10,900"	Mancos/Dolores

ACCESS: Travel north of Mancos on Colo. Hwy. 184 for .2 mi. and turn east onto West Mancos Road (FDR 561). Follow signs 10 miles to Transfer Campground. The trailhead is on the right, directly across from campground. **NARRATIVE:** This trail offers spectacular views of Hesperus Peak as it passes through aspen, spruce and fir forests. Motorized vehicles are not allowed. The trail follows West Mancos Canyon upstream for 5 miles to the historic mining town of Golconda. It continues up the north side of the south fork of the West Mancos River. It meets and follows the Owens Basin Trail for .2 mi., veering left at sign. Keep to the left and follow the switchbacks uphill, heading north to Horsefly Flats. From there the trail follows an old logging road through an open area to the north fork of the West Mancos River and up the drainge to Sharkstooth Trailhead. **USE:** Light. **ACTIVITIES:** HIKING, MTN. BIKES, HORSES. **USGS:** La Plata, Rampart Hills Quads. **MAPS:** 9.

Trail No.	Trail Name	Map Loc.	Distance	Difficulty	Beginning Elev.	Ending Elev.	Ranger District
607 B	**Highline Loop Nat. Rec.**	H 5	17 Mi.	Mod/Diff.	11,600'		Mancos/Dolores

ACCESS: Travel 12 miles north of Hesperus on Cty. Rd. 124. Go on FDR 498 about 2 miles to FDR 571 to Kennebec Pass Trailhead. Hesperus is located about 17 miles east of Mancos or 10 miles west of Durango on U. S.160. Drive about 1/2 mile northwest of Hesperus on U. 5. 160 to reach the beginning of Cty. Rd. 124. **NARRATIVE:** This trail was designated a National Recreation Trail in 1979. The trail was so designated because of its scenic recreational values. Elevations vary from 9,200 feet in the Bear Creek drainage to over 12,000 feet along the "Highline". Aspen, spruce-fir and alpine areas are traversed by the trail as are scattered mountain parks. The beginning of the trail is at Kennebec Pass. Four wheel drive is recommended to reach the pass. The pass is located about 14 miles north of Hesperus on FDR 571. This access begins the traveler at nearly 12, 000 feet elevation. It is then necessary to descend to 9200 feet on the other side of the Highline Divide into Bear Creek and then climb back out to the point of beginning. Other access points are available which require travel along other trails to reach the Highline Loop. One access that can be reached by passenger car is via the Gold Run and Bear Creek Trails. This trail is open for use by horses and hikers. Portions of the trail can also be used by trail bikes. **USE:** Light. **ACTIVITIES:** HIKING, HORSES. **USGS:** La Plata, Orphan Butte, Wallace Ranch Quads. **MAPS:** 5&9.

Trail No.	Trail Name	Map Loc.	Distance	Difficulty	Beginning Elev.	Ending Elev.	Ranger District
607C	**Sharkstooth**	H 5	4.3Mi.	Moderate	10,900'	11,000'	Mancos/Dolores

ACCESS: From Colo. 184 go north on Cty Rd S to FDR 561. Travel northeast on FDR 561 14 miles to the intersection of FDR 561 and Spruce Mill Road (FDR 350) take the Spruce Mill Road 4.5 miles to the intersection of FDR 350 and FDR 346 (Windy Gap Road). Follow the signs to the trailhead. Total mileage is 20 miles. **NARRATIVE:** The trail begins about 1/2 mile beyond Twin Lakes at 10,900 feet and stays to the south of Sharktooth Peak, elevation 11,936 feet. It then descends back into Bear Creek and ends at an elevation of 11,000 feet. The trail joins the Highline National Recreational Trail at Bear Creek. The trail is open to hikers and horses. No motorized vehicles are permitted on trail. **USE:** Moderate. **ACTIVITIES:** HIKING, HORSES. **USGS:** La Plata Quad. **MAP:** 9.

Map No.	Name	Fee	No. of Units	Max. Length	Elev.	Toilets	Water	Ranger District
	CAMPGROUNDS LOCATED IN AREA 1 MAP 9.							
1.	Transfer	$	12	45'	8,500'	Yes	Yes	Mancos/Dolores
2.	Snowslide	$	12	Short	9,000'	Yes	No	Columbine

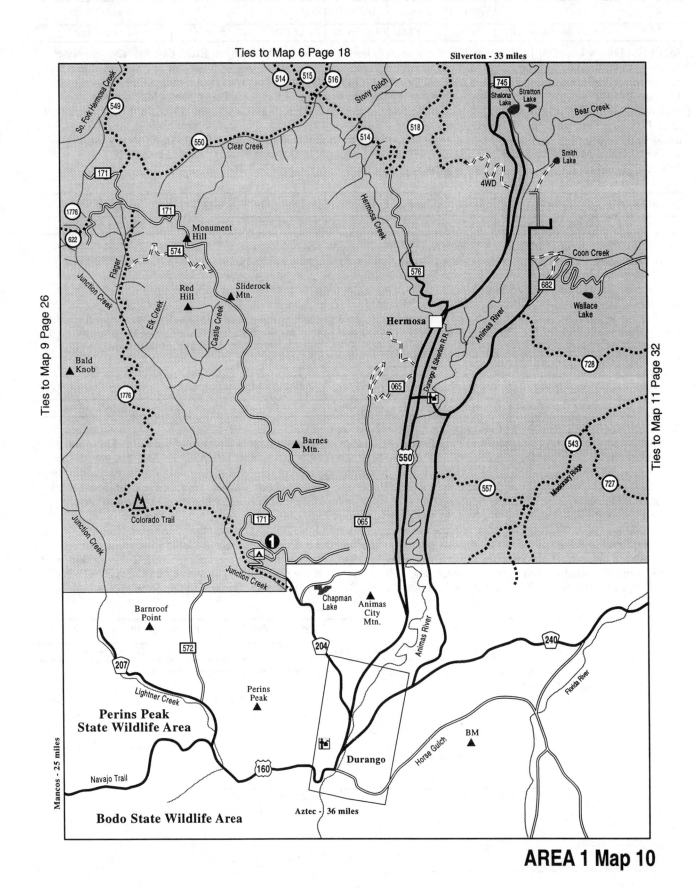

Ties to Map 6 Page 18

Silverton - 33 miles

Ties to Map 9 Page 26

Ties to Map 11 Page 32

So. Fork Hermosa Creek

549

550

Clear Creek

171

1776

622

574

Monument Hill

Red Hill

Sliderock Mtn.

Flager

Junction Creek

Elk Creek

Castle Creek

Bald Knob

1776

Barnes Mtn.

Colorado Trail

Junction Creek

171

Junction Creek

514

515

516

Stony Gulch

514

518

Hermosa Creek

576

745

Shalona Lake

Stratton Lake

Bear Creek

Smith Lake

4WD

Coon Creek

682

Wallace Lake

Hermosa

065

Durango & Silverton R.R.

Animas River

728

550

543

557

Missionary Ridge

727

065

Barnroof Point

572

207

Lightner Creek

Perins Peak

Perins Peak State Wildlife Area

Mancos - 25 miles

Navajo Trail

Chapman Lake

Animas City Mtn.

204

240

Animas River

Florida River

BM

Horse Gulch

160

Durango

Bodo State Wildlife Area

Aztec - 36 miles

Trail No.	Trail Name	Map Loc.	Distance	Difficulty	Beginning Elev.	Ending Elev.	Ranger District
514	**Hermosa Creek**	K 5	18 Mi.	Easy/Mod.	8,400'	7,720'	Columbine

ACCESS: Follow U.S. 550 north 27 miles from Durango to Hermosa Park Road turn-off at Purgatory Ski Area. Take FDR 578 (Hermosa Park Road) to trailhead located by junction of east and north forks of Hermosa Creek, 9 1/2 miles from U.S. 550. **NARRATIVE:** One of the most popular mountain bike routes near Durango. Please expect other cyclists as well as other users including hikers, horses, motorcycles and ATV's. Usually done "downhill" from north to south, the trail loses 1300 feet but regains several hundred feet south of Dutch Creek (Bridged). In some years, the south end can be ridden earlier than many higher trails, but as a round trip retracing the same trail up and back. **USE:** Moderate. **ACTIVITIES:** MOUNTAIN BIKES, HIKING, HORSES, MOTORCYCLES, ATV'S. **USGS:** Monument Hill, Hermosa, Elk Creek Quads. **MAPS:** 6&10.

Trail No.	Trail Name	Map Loc.	Distance	Difficulty	Beginning Elev.	Ending Elev.	Ranger District
518	**Jones Creek**	K 5	3.75 Mi.	More Diff.	7,700'	9,300'	Columbine

ACCESS #1: Take Hwy. 550 north from Durango 10 miles to Hermosa. Turn west on County Road 201 and keep to the right. The road will become FDR 576. Follow the road for approximately 3 miles. The trailhead is on the northside of road where you will find adequate parking. **ACCESS #2:** South portion of Pinkerton-Flagstaff Trail (FDT 522). **ATTRACTIONS:** Jones Creek Trail has beautiful views of the La Platas seen through tall aspen forests. Attractive meadows near top provide plenty of feed for horses and nice campsites with easy access to the creek. Jones Creek Trail is steep in places and winds up through various transitional zones starting in the big Ponderosas, at the upper limit of the oak brush and ending in the aspen and spruce. This provides for an enjoyable variety of sights within a fairly short distance. **USE:** Light. **ACTIVITIES:** HIKING, HORSES. **USGS:** Hermosa, Electra Quads. **MAPS:** 6&10

Trail No.	Trail Name	Map Loc.	Distance	Difficulty	Beginning Elev.	Ending Elev.	Ranger District
549	**South Fork**	J 5	10 Mi.	More Diff.	10,200'	7,800'	Columbine

ACCESS: (Junction Creek) Turn west on 25th St. in Durango. Stay on 25th St. until it turns into Junction Creek Road. Follow for 3 miles until it turns to gravel at the cattle guard. Continue on the main road for approximately 19 miles until you reach a gate, where there is plenty of parking. Hike down the road beyond the gate for about 1-1/2 miles. The road makes a sharp right. Trail starts on the left (southside of the road). **ATTRACTIONS:** This trail provides beautiful meadows, good fishing, great views and access to other trails. **NARRATIVE:** There are a few creek crossings and in the spring it can be a challenge to cross. From the junction, with the Hermosa Trail one could hike down the trail approximately 1 1/2 to 2 miles to Clear Creek and hike back up to the top or you could hike out on the main trail either north or south. **USE:** Light. **ACTIVITIES:** HIKING, FISHING. **USGS:** Monument Hill, Elk Creek Quads. **MAPS:** 6&10.

Trail No.	Trail Name	Map Loc.	Distance	Difficulty	Beginning Elev.	Ending Elev.	Ranger District
550	**Clear Creek**	J 5	6 Mi.	More Diff.	10,200'	7,400'	Columbine

ACCESS: Heading north on Main Street in Durango, turn left (west) on 25th Street until it turns into Junction Creek Road. Follow this road for approximately 22 miles. Trailhead is on the northside of the road. **ATTRACTIONS:** Dropping 2,800 feet from Junction Creek Road with very little uphill in between, this trail is all downhill! Crossing Hermosa Creek at bottom hazardous during spring runoff. It basically runs next to the creek so there is ample water although it should always be filtered to avoid Giardia. Fishing is also a possibility. There are several camp spots along the way especially near the bottom of the trail. There are bears in this area, so caution should be used. **USE:** Medium. **ACTIVITIES:** HIKING, FISHING. **USGS:** Monument Hill Quad. **MAP:** 10.

Map No.	Name	Fee	No. of Units	Max. Length	Elev.	Toilets	Water	Ranger District
CAMPGROUNDS LOCATED IN AREA 1 MAP 10.								
1.	Junction Creek	$	38	50'	7,500'	Yes	Yes	Columbine

AREA 1

Trail No.	Trail Name	Map Loc.	Distance	Difficulty	Beginning Elev.	Ending Elev.	Ranger District
557	**Haflin Creek**	K 5	5 Mi.	More Diff.	6,600'	9,600'	Columbine

ACCESS : Approx. 7 miles north of Durango on East Animas Road (C.R. 250) the trail leads up the eastern side of the valley. Limited parking is available at the trailhead. **ATTRACTIONS:** If using access #1, the trailhead is a short drive from Durango. The trail is usually free of snow by mid to late May. Haflin Trail provides the opportunity to observe a wide variety of tree species including pinion, aspen, gambel oak, ponderosa pine, juniper, fir, and spruce as one passes through the different vegetative zones. The red and white cliffs which the trail follows is used by numerous species of birds as nesting areas. The La Plata mountains and the Hermosa valley are viewable along most of the trail, providing endless photographic opportunities. Haflin Creek is intermittent and is not a reliable source of water. Be sure to plan accordingly. **USE:** Medium. **ACTIVITIES:** HIKING. **USGS:** Durango East Quad. **MAP:** 10.

Trail No.	Trail Name	Map Loc.	Distance	Difficulty	Beginning Elev.	Ending Elev.	Ranger District
622	**Sliderock**	J 5	2.2 Mi.	More Diff.	10,400'	11,800'	Columbine

ACCESS #1: Turn west on 25th St. in Durango. Stay on 25th until it turns into Junction Creek Rd. (FDR 171), and continue to follow it 3 miles until it crosses the cattle guard and turns into dirt. Follow this road (FDR 171), for approx. 17.5 miles. Take FDR 171 north to Champion Venture Rd. and go 0.7 miles to where the trail crosses the road. Limited parking is available. **ACCESS #2:** Drive west on U.S. 160 from Durango for 15 miles to the La Plata Canyon turn-off (FDR 571). Follow this road for 1 mile to Kennebec Pass. 4WD with good clearance is needed for the last section of the road. Ample parking is available at the trailhead. **ATTRACTIONS:** The Sliderock Trail is a small portion of the Colorado Trail. This trail offers excellent views and photographic opportunities, especially on Kennebec Pass. A consideration is the exposure on this trail, especially in the areas at or above timberline. High country storms can move in quickly, so be prepared for inclement weather. No water is available on this trail. The first section of trail climbs up several switchbacks, then eases around the ridge. The trail then breaks out of the timber and cuts upward across a sloped field. This trail also offers access to Cumberland Mountain. **USE:** Heavy. **ACTIVITIES:** HIKING. **USGS:** La Plata, Monument Hill Quads. **MAP:** 10.

Trail No.	Trail Name	Map Loc.	Distance	Difficulty	Beginning Elev.	Ending Elev.	Ranger District
727	**First Fork**	L 6	3.4 Mi.	More Diff.	7,800'	9,500"	Columbine

ACCESS #1: Drive approx. 8 miles east of Durango on Florida Road, County Road 240. Turn north at Colvig Silver Camp and go past the camp, over the cattle guard and continue to follow the road for approx. 1 mile. Strong 2WD with good clearance may be needed for the last mile of the road. Limited parking is available. **ACCESS #2:** Via Missionary Ridge Trail (FDT 543). **ATTRACTIONS:** This trail is excellent during late spring through mid-fall due to its relatively low elevation and southeast facing slopes. The trail winds through timber on the lower section, but opens up further along the trail. The trail follows a stream, which is crossed several times. Water availability is not a problem, but all water should be purified due to problems with Giardia. This trail is excellent for viewing the fall colors of the gambel oak and aspen trees. It is not uncommon to see large game such as elk and mule deer in the First Fork/Missionary Ridge Trail areas. **USE:** Light. **ACTIVITIES:** HIKING. **USGS:** Rules Hill, Durango East, Hermosa Quads. **MAPS:** 10&11.

CODE	NATIONAL FOREST	PHONE NO.
ARNF	APAPAHO/ROOSEVELT	
BRD	BOULDER RANGER DISTRICT	(303) 444-6600
CCRD	CLEAR CREEK RANGER DISTRICT	(303) 567-2901
EPRD	ESTES-POUDRE RANGER DISTRICT	(970) 498-2770
PWRD	PAWNEE RANGER DISTRICT	(970) 353-5004
RFRD	RED FEATHER RANGER DISTRICT	(970) 498-2770
SURD	SULPHUR RANGER DISTRICT	(970) 887-4100
GMNF	GRAND MESA	
CBRD	COLLBRAN RANGER DISTRICT	(970) 487-3534
GJRD	GRAND JUNCTION RAN. DISTRICT	(970) 242-8211
GNF	GUNNISON	
CERD	CEBOLLA RANGER DISTRICT	(970) 641-0471
PRD	PAONIA RANGER DISTRICT	(970) 527-4131
TRRD	TAYLOR RIVER RANGER DISTRICT	(970) 641-0471
PNF	PIKE	
PPRD	PIKES PEAK RANGER DISTRICT	(719) 636-1602
SKRD	SOUTH PARK RANGER DISTRICT	(719) 836-2031
SPRD	SOUTH PLATTE RANGER DISTRICT	(303) 275-5610
RGNF	RIO GRANDE/SAN JUAN	
CPRD	CONEJOS PEAK RANGER DISTRICT	(719) 274-8971
CRRD	DIVIDE/CREEDE RANGER DISTRICT	(719) 658-2556
DNRD	DIVIDE/DEL NORTE RANGER DIST.	(719) 657-3321
SARD	SAGUACHE RANGER DISTRICT	(719) 655-2547
RNF	ROUTT	
HBRD	HAHNS PEAK/BEAR EARS RANGER DIST.	(970) 879-1870
PWRD	THE PARKS RANGER DISTRICT	(970) 723-8204
YRD	YAMPA RANGER DISTRICT	(970) 638-4516
SINF	SAN ISABEL	
LRD	LEADVILLE RANGER DISTRICT	(719) 486-0749
SRD	SALIDA RANGER DISTRICT	(719) 539-3591
SCRD	SAN CARLOS RANGER DISTRICT	(719) 269-8500
SJNF	SAN JUAN/RIO GRANDE	
CBRD	COLUMBINE RANGER DIST.(BAYFIELD)	(970) 884-2512
CRD	COLUMBINE RANGER DIST. (DURANGO)	(970) 247-4874
DMRD	DOLORES/MANCOS RANGER DIST.	(970) 882-7296
MDRD	MANCOS/DOLORES RANGER DIST.	(970) 533-7716
PARD	PAGOSA RANGER DISTRICT	(970) 264-2268
UNCNF	UNCOMPAHGRE	
GJRD	GRAND JUNCTION RANGER DIST.	(970) 242-8211
NRD	NORWOOD RANGER DISTRICT	(970) 327-4261
ORD	OURAY RANGER DISTRICT	(970) 240-5300
WRNF	WHITE RIVER	
ARD	ASPEN RANGER DISTRICT	(970) 925-3445
BLRD	BLANCO RANGER DISTRICT	(970) 878-4039
DRD	DILLON RANGER DISTRICT	(970) 468-5400
ERD	EAGLE RANGER DISTRICT	(970) 328-6388
HCRD	HOLY CROSS RANGER DISTRICT	(970) 827-5715
RRD	RIFLE RANGER DISTRICT	(970) 625-2371
SORD	SOPRIS RANGER DISTRICT	(970) 963-2266
	OTHER AGENCIES	
BLM	BUREAU OF LAND MANAGEMENT	
GJ	GRAND JUNCTION RESOURCE AREA	(970) 244-3000
GUN	GUNNISON RESOURCE AREA	(970) 641-0471
GWS	GLENWOOD SPRINGS RES. AREA	(970) 947-2800
LSK	LITTLE SNAKE RESOURCE AREA	(970) 824-4441
SJ	SAN JUAN RESOURCE AREA	(970) 247-4082
SL	SAN LUIS RESOURCE AREA	(719) 589-4975
UNC	UNCOMPAHGRE BASIN RES. AREA	(970) 240-5300
BMP	BOULDER MTN. PARKS	(303) 441-3408

CODE	AGENCY	PHONE NO.
COE	CORP. OF ENGINEERS	
	JOHN MARTIN RESERVOIR	(719) 336-3476
COL	CITY OF LAKEWOOD	
	BEAR CREEK LAKE PARK	(303) 697-6159
COR	CITY OF RIFLE	
	RIFLE MOUNTAIN PARK	(970) 625-2121
COT	CITY OF TRINIDAD	
	MONUMENT LAKE RESORT	(719) 868-2226
DOW	DIVISION OF WILDLIFE	
CEN	CENTRAL REGION, DENVER	(303) 297-1192
NE	NORTHEAST REGION, FT. COLLINS	(970) 484-2836
NW	NORTHWEST REG.GRAND JUNCTION	(970) 284-7175
SE	SOUTHEAST REGION, COLO SPRINGS	(719) 473-2945
SW	SOUTHWEST REGION, MONTROSE	(970) 249-3431
DPOR	DIVISION OF PARKS AND OUTDOOR RECREATION	
MAIN	DIRECTORS OFFICE	(303) 866-3437
METRO	METRO DENVER, LITTLETON	(303) 791-1957
NO	NORTH REGION, FT. COLLINS	(970) 226-6641
SO	SOUTH REGION, COLORADO SPGS	(719) 471-0900
WEST	WEST REGION, CLIFTON	(970) 434-6862
FAWS	U.S. FISH AND WILDLIFE SERVICE	
	BROWNS PARK NAT. WILDLIFE REFUGE	(970) 365-3613
HCO	HINSDALE COUNTY	
	COUNTY OFFICE, LAKE CITY	(970) 944-2225
LCP	LARMIER COUNTY PARKS	
	PARKS OFFICE, FORT COLLINS	(970) 679-4570
LPD	LONGMONT CITY PARKS DEPARTMENT	
	UNION RESERVOIR PARK OFFICE	(303) 772-1265
NPS	NATIONAL PARKS SERVICE	
	BLACK CANYON OF THE GUNNISON	(970) 641-2337
	COLORADO NATIONAL MONUMENT	(970) 641-2337
	CURECANTI NAT. RECREATION AREA	(970) 641-2337
	DINOSAUR NATIONAL MONUMENT	(970) 374-3000
	GREAT SAND DUNES NATIONAL MON.	(719) 378-2312
	MESA VERDE NATIONAL PARK	(970) 529-4465
	ROCKY MOUNTAIN NAT. PARK	(970) 586-1206

CAMPING RESERVATIONS:

NATIONAL FOREST SERVICE	(800) 280-CAMP
STATE PARKS (From outside Metro Denver Area)	(800) 678-2267
(From inside Metro Denver Area)	(303) 470-1144

NOTE: State Parks reservation period April through Sept.

NATIONAL PARKS SERVICE	(800) 365-2267
(Rocky Mountain National Park)	

NOTE:
1. Phone numbers and campground information to 1-1-97.
2. Campground Fees vary and are subject to change.

CAMPGROUND INFORMATION PAGE 34

Ties to Map 7 Page 24

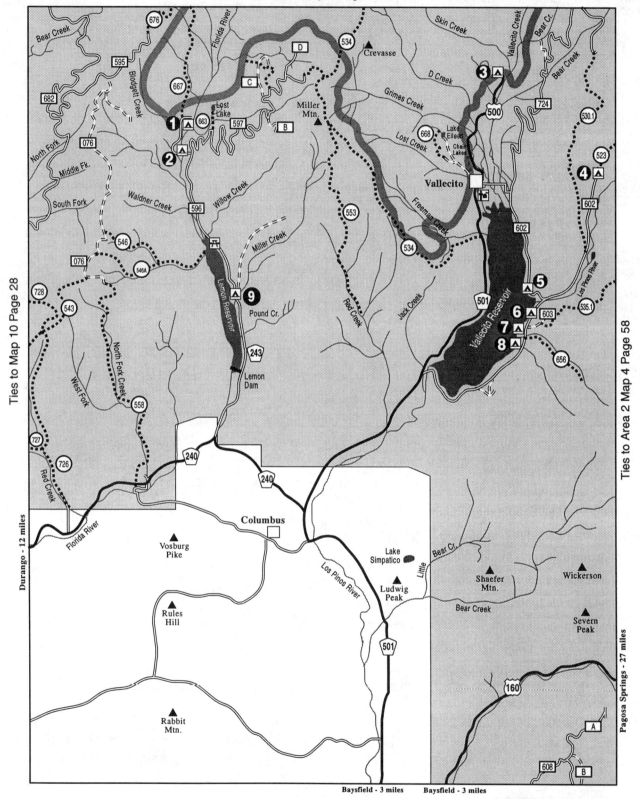

Ties to Map 10 Page 28

Durango - 12 miles

Ties to Area 2 Map 4 Page 58

Pagosa Springs - 27 miles

Baysfield - 3 miles   Baysfield - 3 miles

AREA 1 Map 11

Trail No.	Trail Name	Map Loc.	Distance	Difficulty	Beginning Elev.	Ending Elev.	Ranger District
535	**East Creek**	M 5	9.6 Mi.	Mod/Diff.	7,850'	10,300'	Columbine

ACCESS: From Bayfield Travel 13 miles north on County Road 501 to Vallecito Reservoir. Turn east on FDR 603, cross the dam on the south end of the reservoir. Follow along the the east side of the reservoir for about 3.5 miles. Turn right onto FDR 852, continue .5 mile to trailhead. **NARRATIVE:** The trailhead sign is on the rightside of the road, the trail passes through a gate that should be kept closed. About .5 mile up is a scenic overlook which views Vallecito Reservoir. Motorized vehicles are allowed only for the first 3.75 miles. At about 5.5 miles the trail crosses East Creek and follows the north side of the creek. It joins the Pine-Piedra Stock Driveway Trail at the ridgeline between the Los Pinos and Piedra River watersheds. **USE:** Moderate. **ACTIVITIES:** HIKING, HORSES. **USGS:** Vallecito Reservoir. **MAP:** 11.

Trail No.	Trail Name	Map Loc.	Distance	Difficulty	Beginning Elev.	Ending Elev.	Ranger District
546A	**Youngs Canyon**	L 5	4.5 Mi.	Most Diff.	8,100'	10,200'	Columbine

ACCESS #1: Drive east out of Durango on Florida Road (C.R. 240) approximately 12 miles to the junction with C.R. 243. Go north on C.R. 243 to the north (upper) end of Lemon Reservoir. Forest Service public parking with facilities is available. For public access to the trail, wading the Florida River is necessary. However, there is a private bridge across the river about 1/2 mile north of the parking area. Be sure to ask permission before crossing private land. **ACCESS #2:** This trail access is on the Missionary Ridge Road (FDR 076), near Horse Thief Park. **ATTRACTIONS:** The trail contours uphill all the way to Horse Thief Park on the top of Missionary Ridge. Also, the trail stays on mostly southern slopes so it may be hikeable in early spring. This trail can be accessed by hiking past the upper part of the lake and wading the stream. **NARRATIVE:** The trail is very steep for the first 1.5 miles with about 1,200 foot rise in elevation. Trail is heavily used during the hunting seasons. **USE:** Light. **ACTIVITIES:** HIKING. **USGS:** Lemon Reservoir Quad. **MAP:** 11.

Trail No.	Trail Name	Map Loc.	Distance	Difficulty	Beginning Elev.	Ending Elev.	Ranger District
558	**Shearer Creek**	L 5	7.7 Mi.	More Diff.	7,600'	10,000'	Columbine

ACCESS: Follow Florida Road approximately 9 1/2 miles east of Durango to a dirt pull-off on the left side of the road directly across from the restuarant parking lot. A few parking spots are available and more is available across the road. **ATTRACTIONS:** This trail is long and hot in the summer making it a nice spring hike. A couple of stream crossings are necessary near the bottom of the trail. This is a heavily used horse trail so caution should be used and water purified to avoid Giardia. There are beautiful views along the top of the trial. **USE:** Heavy. **ACTIVITIES:** HIKING, HORSES. **USGS:** Rules Hill, Lemon Reservoir Quads. **MAP:** 11.

Trail No.	Trail Name	Map Loc.	Distance	Difficulty	Beginning Elev.	Ending Elev.	Ranger District
656	**North Canyon**	M 5	1.6 Mi.	More Diff	7,730'	8,800'	Columbine

ACCESS: The trail can be accessed by traveling 3.5 miles to FDR 603 which begins below the Vallecito Dam off of County Road 501. Cross the dam and travel along the east side of the reservoir. The trail begins on the right side of the road near the entrance to North Canyon Campground. **ATTRACTIONS:** Scenery. Special features: horse use, foot use and mountain bikes. **USE:** Moderate. **ACTIVITIES:** HIKING, MOUNTAIN BIKES, HORSES. **USGS:** Vallecito Reservoir Quad. **MAP:** 11.

Trail No.	Trail Name	Map Loc.	Distance	Difficulty	Beginning Elev.	Ending Elev.	Ranger District
663	**Lost Lake**	L 5	0.5 Mi.	Easy	8,900'	9,100'	Columbine

ACCESS: This trail can be reached by driving north of Lemon Reservoir on FDR 596. You will need to drive through 2 miles of private land after which you will meet FDR 597 or East Florida Road. Drive 1 3/4 miles up this road. The trail begins on the left side of the road on a switchback. Parking is available on the roadside. **ATTRACTIONS:** The trail ends at Lost Lake. This is a very small, shallow lake which may or may not have fish in it. Very scenic. **USE:** Moderate. **ACTIVITIES:** HIKING, FISHING. **USGS:** Lemon Reservoir Quad. **MAP:** 11.

Trail No.	Trail Name	Map Loc.	Distance	Difficulty	Beginning Elev.	Ending Elev.	Ranger District
726	**Red Creek**	L 6	3.5 Mi.	Most Diff.	8,000'	9,800'	Columbine

ACCESS #1: From Durango take County Road 240, follow Florida Road 8 miles east and turn north at Colvig Silver Camp. Go past the camp, over the cattle guard and follow the dirt road approx. 1-1/2 miles to the trailhead, which is at the end of this road. Strong 2WD or 4WD with good clearance may be needed for the last portion of the dirt road. **ACCESS #2:** Via the Missionary Ridge Trail (FDT 543). **ATTRACTIONS:** Like the First Fork Trail, this trail provides good hiking in the late spring through the mid-fall. The trail stays next to the stream most of the way up in the shade of big fir, spruce and aspen trees. The trail gradually climbs the first 2-1/2 miles and then goes up a series of switchbacks which meets up with the Missionary Ridge Trail. As with nearly all water in the forest, it is advisable to use a filter to prevent problems with Giardia. Fall colors on this trail are spectacular and sightings of large game animals such as elk and mule deer are not uncommon. An overnight loop trip utilizing Missionary Ridge Trail and First Fork Trail is possible. **USE:** Heavy. **ACTIVITIES:** HIKING. **USGS:** Rules Hill, Lemon Reservoir Quads. **MAP:** 11.

Trail No.	Trail Name	Map Loc.	Distance	Difficulty	Beginning Elev.	Ending Elev.	Ranger District
668	**Lake Eileen**	M 5	1.5 Mi.	More Diff.	7,720'	8,860'	Columbine

ACCESS: The trail can be accessed by driving past the north end of Vallecito Reservoir on County Road 501 to the Forest Service Work Center. Park on the side of the road just beyond the work center entrance. The trail signs can be seen on each side of the road. If no parking is available at the signs, continue north around the corner for 500 feet to a small pull-off on the right. **ATTRACTIONS:** The trail is clearly marked. It is quite steep in places. The small shallow lake lies in a small depression surrounded by aspens. The lake is covered by water lilies and is especially pretty when these bloom. There are no fish in the lake. Horses seldom use this trail. **USE:** Moderate. **ACTIVITIES:** HIKING. **USGS:** Vallecito Reservoir Quad. **MAP:** 11.

Map No.	Name	Fee	No. of Units	Max. Length	Elev.	Toilets	Water	Ranger District
	CAMPGROUNDS LOCATED IN AREA 1 MAP 11. (Page 32)							
1.	Transfer Park	$	25	35'	8,600'	Yes	Yes	Columbine
2.	Florida & Florida Group	$	20	35'	8,500'	Yes	Yes	Columbine
3.	Vallecito	$	80	35'	8,000'	Yes	Yes	Columbine
4.	Pine River	$	6	15'	8,100'	Yes	No	Columbine (Best For Tents)
5.	Middle Mountain	$	24	35'	7,900'	Yes	Yes	Columbine
6.	Pine Point	$	30	35'	7,900'	Yes	Yes	Columbine
7.	North Canyon	$	21	35'	7,900'	Yes	Yes	Columbine
8.	Graham Creek	$	25	35'	7,900'	Yes	Yes	Columbine
9.	Miller Creek	$	12	35'	8,000'	Yes	Yes	Columbine

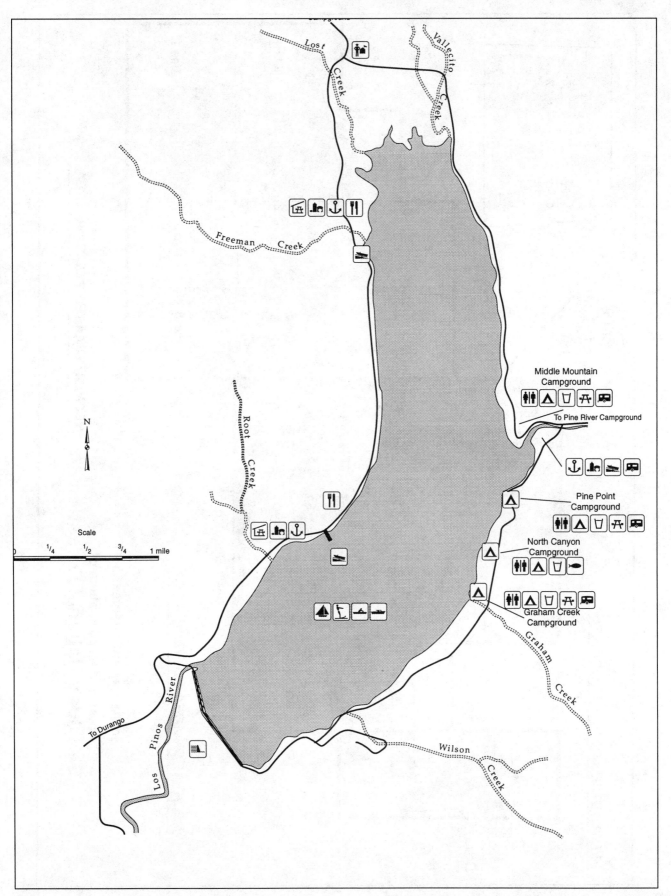

Lost Creek

Vallecito Creek

Freeman Creek

Root Creek

N

Scale

1/4   1/2   3/4   1 mile

Middle Mountain
Campground

To Pine River Campground

Pine Point
Campground

North Canyon
Campground

Graham Creek
Campground

Graham Creek

To Durango

Los Pinos River

Wilson Creek

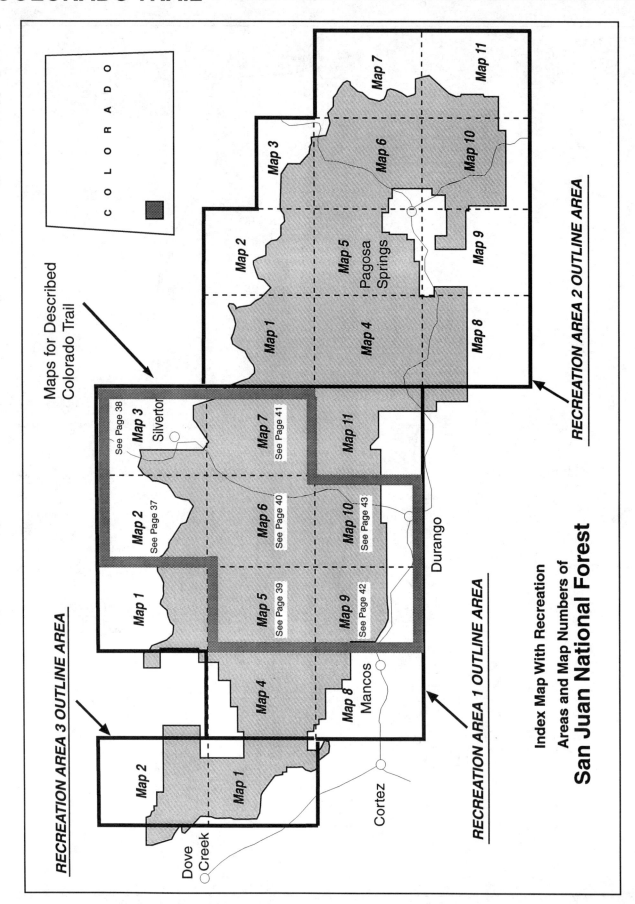

Maps for Described
Colorado Trail

RECREATION AREA 3 OUTLINE AREA

RECREATION AREA 1 OUTLINE AREA

RECREATION AREA 2 OUTLINE AREA

COLORADO

Index Map With Recreation
Areas and Map Numbers of
San Juan National Forest

Map 7
Map 11
Map 3
Map 6
Map 10
Map 2
Map 5
Map 9
Map 1
Map 4
Map 8

Pagosa
Springs

Map 3
Silverton
See Page 38

Map 7
See Page 41

Map 11

Map 2
See Page 37

Map 6
See Page 40

Map 10
See Page 43

Map 1

Map 5
See Page 39

Map 9
See Page 42

Durango

Map 2

Map 1

Map 4

Map 8
Mancos

Cortez

Dove
Creek

TRAIL NARRATIVE PAGE 44

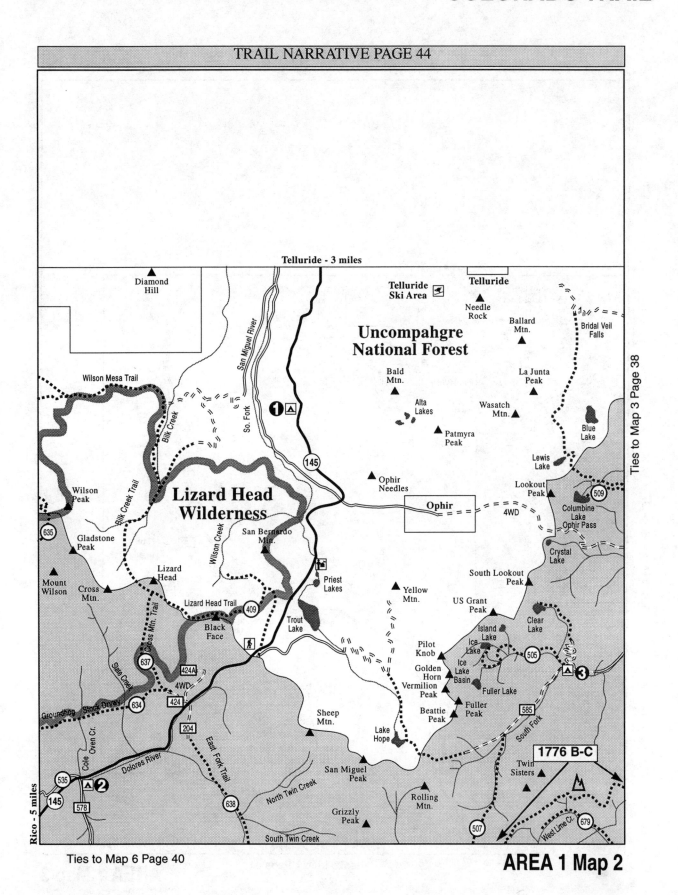

Telluride - 3 miles

Diamond Hill

Telluride Ski Area

Telluride

Needle Rock

Ballard Mtn.

Bridal Veil Falls

Wilson Mesa Trail

San Miguel River

So. Fork

Uncompahgre National Forest

Bald Mtn.

La Junta Peak

Bilk Creek

Alta Lakes

Wasatch Mtn.

Blue Lake

Bilk Creek Trail

①

Patmyra Peak

Lewis Lake

Wilson Peak

Lizard Head Wilderness

145

Ophir Needles

Lookout Peak

509

Columbine Lake
Ophir Pass

Ophir

4WD

Gladstone Peak

635

San Bernardo Mtn.

Wilson Creek

Crystal Lake

Mount Wilson

Lizard Head

Priest Lakes

South Lookout Peak

Cross Mtn.

Yellow Mtn.

Lizard Head Trail

409

US Grant Peak

Clear Lake

Cross Mtn. Trail

Black Face

Trout Lake

Island Lake

Ice Lake

505

③

637

424A

Pilot Knob

Ice Lake Basin

Fuller Lake

Golden Horn

Slate Creek

4WD

Vermilion Peak

Fuller Peak

Groundhog

Stock Driveway

634

424

Beattie Peak

585

South Fork

Cole Oven Cr.

204

Sheep Mtn.

Lake Hope

1776 B-C

Dolores River

East Fork Trail

Twin Sisters

535

②

San Miguel Peak

145

578

638

North Twin Creek

Rolling Mtn.

507

West Lime Cr.

679

Grizzly Peak

Rico - 5 miles

South Twin Creek

Ties to Map 6 Page 40

Ties to Map 3 Page 38

AREA 1 Map 2

COLORADO TRAIL

Ouray - 7 miles

Telluride Peak

Uncompahgre National Forest

Ptarmigan Lake

Turtle Mtn.

Cinnamon Mtn.

Lake Como

California Mtn.

Red Mtn.

Red Mtn. No. 1

Hurricane Peak

Treasure Mtn.

Red Mtn. Pass

Red Mtn. No. 3

Hanson Peak

823

4WD

Lake Emma

Bonita Peak

Eureka Mtn.

Niagara Peak

McMillan Peak

Crown Mtn.

Bullion King Lake

Cement Creek

Emery Peak

Browns Gulch

Chattanooga

509

Dome Mtn.

825

Storm Peak

Tower Mtn.

679

4WD

Ohio Peak

Middle Mtn.

588

Niagara Gulch

50

Macomber Peak

Howardsville

586

Galena Mtn.

Crystal Lake

585

110

Animas River

Cunningham Creek

589

4WD

Silverton

737

Canby Mtn.

VABM

Kendall Mtn.

Hazelton Mtn.

Bear Creek

Bear Mtn.

Sultan Mtn.

Kendall Peak

Silver Lake

Green Mtn.

502

1776 B-C

Grand Turk

550

674

Sugarloaf

606

Weminuche Wilderness

Highland Mary Lakes

Deep Cr.

Mabel Mine

Whitehead

813

Verde Lakes

813

Rio Grand Nat'l. Forest

Durango - 42 miles

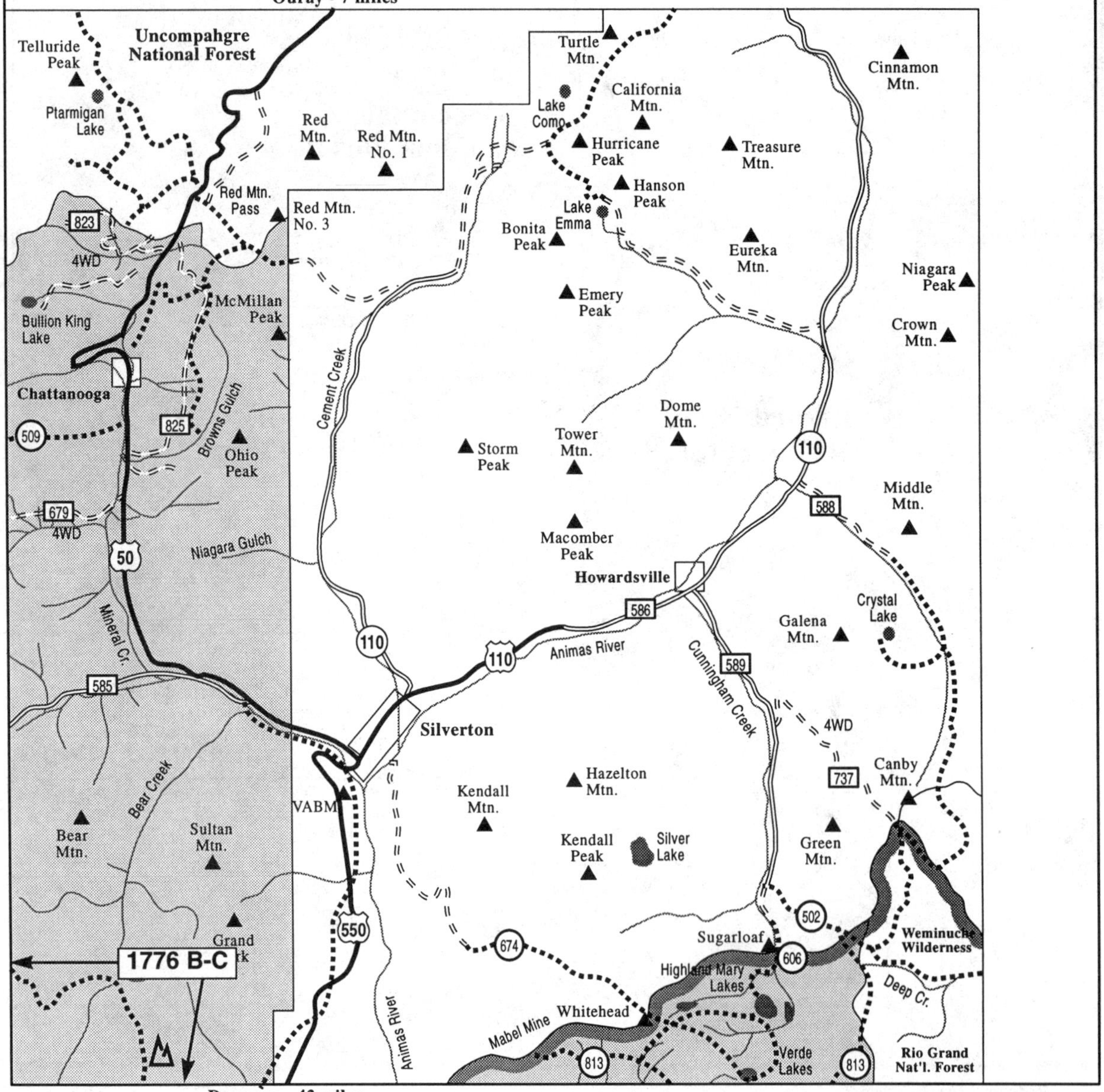

Ties to Map 2 Page 37

Ties to Map 7 Page 41

AREA 1 Map 3

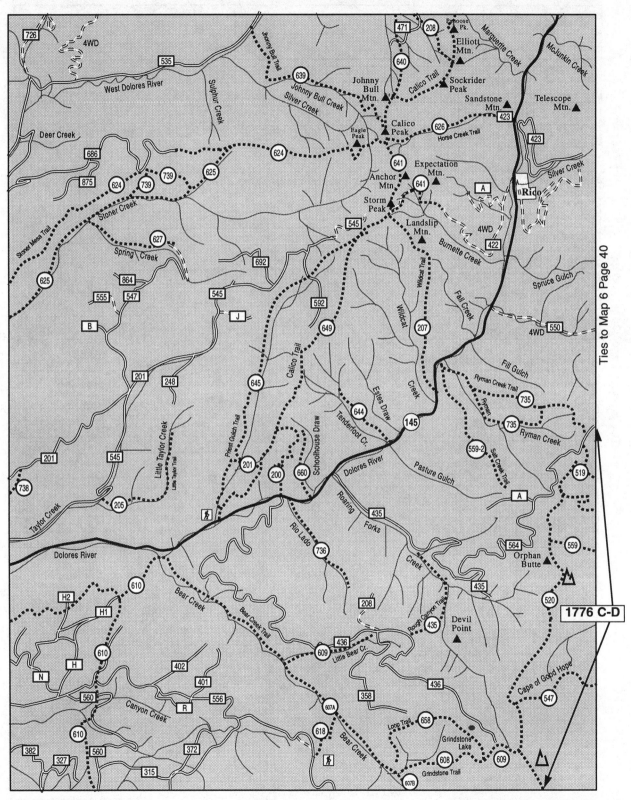

Ties to Map 6 Page 40

1776 C-D

AREA 1 Map 5

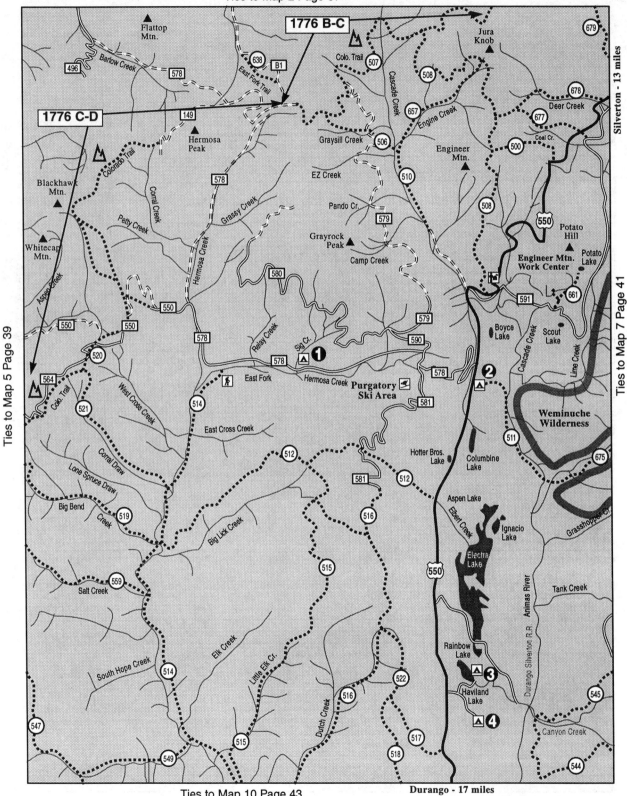

TRAIL NARRATIVE PAGE 44

Ties to Map 2 Page 37

1776 B-C

1776 C-D

Ties to Map 5 Page 39

Ties to Map 7 Page 41

Silverton - 13 miles

Flattop Mtn.

496   Barlow Creek   578   638   B1   East Fork Trail   Colo. Trail   507   508   Jura Knob   679

149   Hermosa Peak   Graysill Creek   506   657   Engine Creek   678   Deer Creek   677   500   Coal Cr.

Blackhawk Mtn.   Colorado Trail   578   EZ Creek   510   Engineer Mtn.   508   550   Potato Hill

Whitecap Mtn.   Corral Creek   Petty Creek   Pando Cr.   579   Engineer Mtn. Work Center   Potato Lake

Aspen Creek   Grassy Creek   Hermosa Creek   Grayrock Peak   Camp Creek   580   591   661   Scout Lake

550   550   579   Boyce Lake   Cascade Creek   Lime Creek

550   578   Relay Creek   Sig Cr.   590   ⚑1   578   2⚑   Weminuche Wilderness

520   East Fork   Hermosa Creek   Purgatory Ski Area   581   511   675

564   Colo. Trail   514   East Cross Creek   Hotter Bros. Lake   Columbine Lake

521   West Cross Creek   512   581   512   Aspen Lake   Ignacio Lake   Grasshopper Cr.

Corral Draw   Lone Spruce Draw   516   Electra Lake   Animas River   Tank Creek

Big Bend   519   Creek   Big Lick Creek   515   550   Durango Silverton R.R.

559   Salt Creek   Rainbow Lake   3⚑

547   South Hope Creek   514   Elk Creek   Little Elk Cr.   Dutch Creek   516   522   Haviland Lake   4⚑   545   Canyon Creek

549   515   517   518   544

Ties to Map 10 Page 43

Durango - 17 miles

AREA 1 Map 6

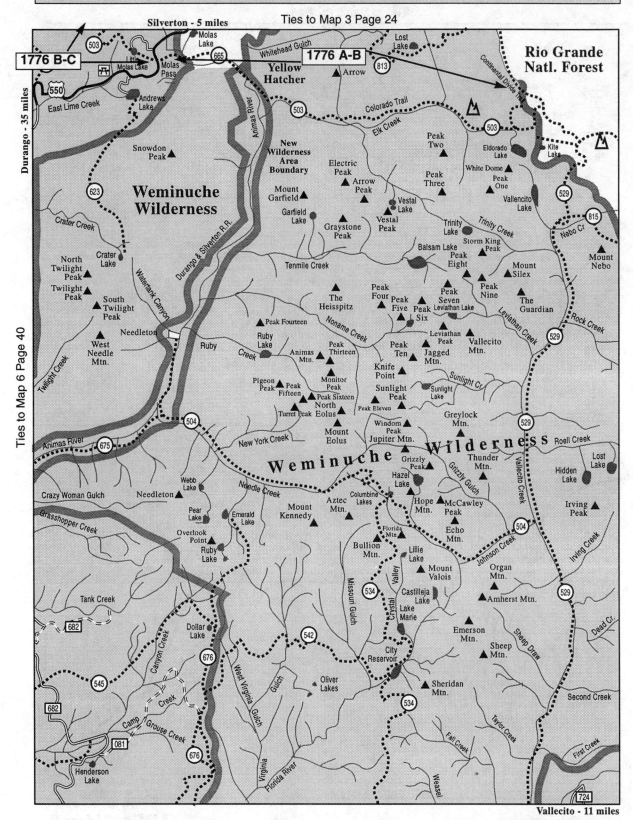

Silverton - 5 miles

1776 B-C

1776 A-B

Rio Grande
Natl. Forest

503

Molas Lake
665
Whitehead Gulch
813
Lost Lake

550
Little Molas Lake
Molas Pass
Yellow Hatcher
Arrow
Continental Divide Trail

Durango - 35 miles
East Lime Creek
Andrews Lake
Colorado Trail
503
503
Kite Lake

Snowdon Peak
New Wilderness Area Boundary
Electric Peak
Elk Creek
Peak Two
Eldorado Lake

623
Weminuche Wilderness
Arrow Peak
Mount Garfield
White Dome
Peak One
Vallencito Lake
529

Crater Creek
Garfield Lake
Graystone Peak
Vestal Lake
Vestal Peak
Trinity Lake
Trinity Creek
815

North Twilight Peak
Crater Lake
Tenmile Creek
Balsam Lake
Storm King Peak
Mount Nebo

Twilight Peak
Watertank Canyon
The Heisspitz
Peak Four
Peak Eight
Peak Seven
Mount Silex
Mount Nebo

South Twilight Peak
Durango & Silverton R.R.
Animas River
Noname Creek
Peak Five
Peak Six
Leviathan Lake
Peak Nine
The Guardian
Nebo Cr

Needleton
Peak Fourteen
Ruby Lake
Ruby Creek
Peak Thirteen
Peak Ten
Leviathan Peak
Vallecito Mtn.
Leviathan Creek
Rock Creek
529

West Needle Mtn.
Animas Mtn.
Knife Point
Jagged Mtn.

Ties to Map 6 Page 40
Twilight Creek
Pigeon Peak
Peak Fifteen
Monitor Peak
Sunlight Peak
Sunlight Lake
Sunlight Cr.

Turret Peak
Peak Sixteen
North Eolus
Peak Eleven
Windom Peak
Greylock Mtn.
529

504
Mount Eolus
Jupiter Mtn.
Roell Creek

New York Creek
Weminuche Wilderness
Grizzly Peak
Thunder Mtn.

Animas River
675
Hazel Lake
Grizzly Gulch
Hidden Lake
Lost Lake

Crazy Woman Gulch
Webb Lake
Needleton
Needle Creek
Columbine Lakes
Hope Mtn.
McCawley Peak
Vallecito Creek
Irving Peak

Grasshopper Creek
Pear Lake
Emerald Lake
Aztec Mtn.
Echo Mtn.
504
Irving Creek

Overlook Point
Ruby Lake
Mount Kennedy
Florida Mtn.
Organ Mtn.

Tank Creek
Bullion Mtn.
Lillie Lake
Mount Valois
Amherst Mtn.
529
Dead Cr.

682
Dollar Lake
542
Missouri Gulch
534
Crystal Valley
Castilleja Lake
Lake Marie
Emerson Mtn.
Sheep Draw

676
Canyon Creek
West Virginia Gulch
City Reservoir
Sheep Mtn.
Second Creek

545
Creek
Oliver Lakes
534
Sheridan Mtn.
Taylor Creek

682
Camp Grouse Creek
081
Virginia Gulch
Florida River
Fall Creek
Weasel
First Creek

676
Henderson Lake
724

Vallecito - 11 miles

AREA 1 Map 7

TRAIL NARRATIVE PAGE 45

Ties to Map 5 Page 39

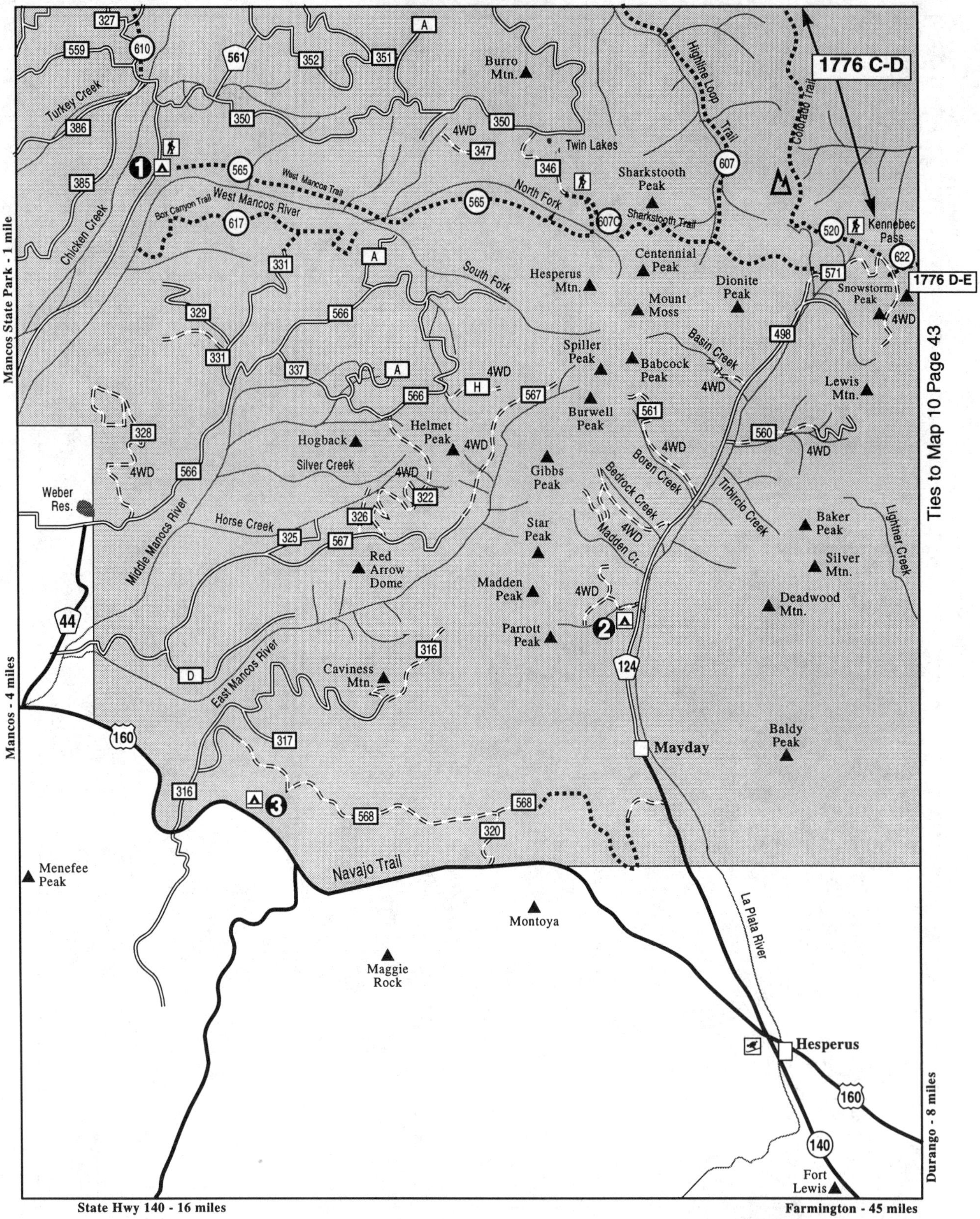

327
559
610
561
352
351
A

Turkey Creek

386
350

Burro Mtn. ▲

Highline Loop Trail

Colorado Trail

1776 C-D

385
4WD 350
347
565
West Mancos Trail
West Mancos River
565
North Fork
346
Sharkstooth Peak ▲

607
607C
Sharkstooth Trail

Kennebec Pass
520
622

1
Box Canyon Trail
617

Mancos State Park - 1 mile
Chicken Creek

331
A
South Fork

Hesperus Mtn. ▲

Centennial Peak ▲

Dionite Peak ▲

571
1776 D-E

Snowstorm Peak ▲
4WD

329
566

Mount Moss ▲

498

Basin Creek

331

Spiller Peak ▲

Babcock Peak ▲

Lewis Mtn. ▲

337
A
4WD
566
567

Burwell Peak ▲

561
4WD
560
4WD

H

Helmet Peak ▲

Hogback ▲
Silver Creek
4WD

Gibbs Peak ▲

Boren Creek
Bedrock Creek
Tirbircio Creek

328
4WD
566

322

Baker Peak ▲

Weber Res.

Horse Creek
325
567
326

Star Peak ▲

Madden Cr.
4WD

Silver Mtn. ▲

Middle Mancos River

Red Arrow Dome ▲

Madden Peak ▲
4WD

Deadwood Mtn. ▲

44

316

2

East Mancos River

Parrott Peak ▲

124

Mancos - 4 miles

D
316

Caviness Mtn. ▲

160
317

3
568
568
320

Baldy Peak ▲

Mayday

Menefee Peak ▲

Navajo Trail

La Plata River

Montoya ▲

Maggie Rock ▲

Hesperus

160

Durango - 8 miles

140

Fort Lewis ▲

State Hwy 140 - 16 miles

Farmington - 45 miles

Ties to Map 10 Page 43

AREA 1 Map 9

Ties to Map 6 Page 40

Silverton - 33 miles

Ties to Map 9 Page 42

Mancos - 25 miles

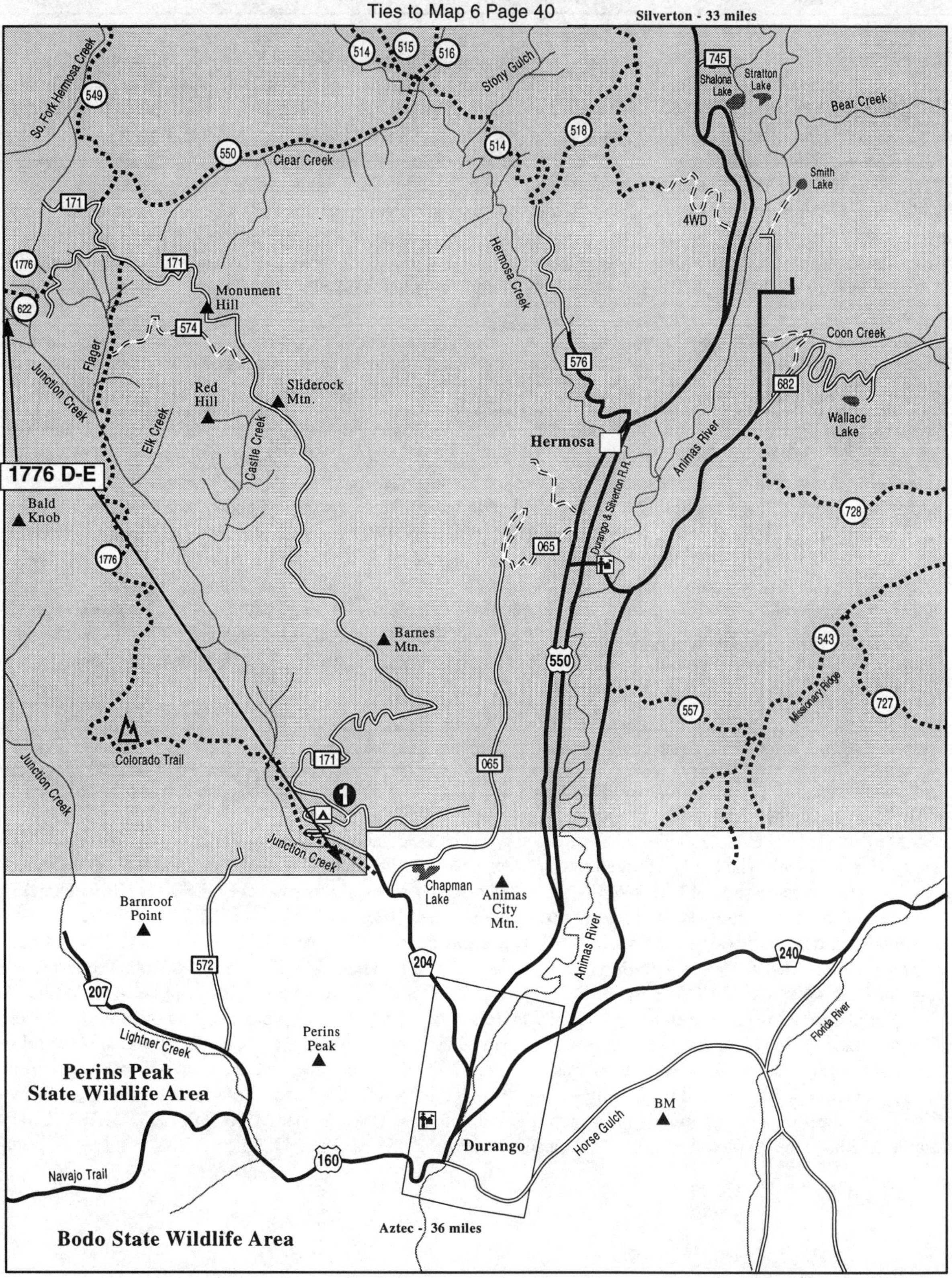

AREA 1 Map 10

COLORADO TRAIL

Trail No.	Trail Name	Map Loc.	Distance	Difficulty	Beginning Elev.	Ending Elev.	Ranger District
1776 A-B	**Colorado**	M 3	12.9 Mi.	Most Diff.	12,680'	10,880'	Columbine

Continental Divide to Molas Pass

ACCESS #1: Go west of Creede on Colorado 149 21 miles, turn left at the Rio Grande Reservoir and continue for 19 miles. The gravel road ends and continues on 4WD road for 7 miles. Start at Pole Creek Trail. The Continental Divide is 8.3 miles. **ACCESS #2:** Go 5.5 miles south of Silverton to Molas Lake-Molas Trail Colorado Trail is .2 miles south of actual trailhead and parking area. **ACCESS #3:** Ride the Durango-Silverton Narrow Gauge Railroad to Elk Park on the Animas River. **ATTRACTIONS:** The Colorado Trail enters the San Juan NF at the Continental Divide and descends on the Elk Creek Trail to the Animas River, then ascends the Molas Trail to US 550. Snowfields in Elk Creek's narrow gorge may linger well into mid-June and campsites are limited at both upper and lower sections of Elk Creek due to the steepness of the canyon walls. However, there is a large park in the mid-section to allow ample camping sites. There is also limited camping on the lower portion of the Molas Trail. Please practice low impact camping due to heavy use in the area. **USE:** Heavy. **ACTIVITIES:** HIKING. **USGS:** Snowdon Park, Storm King Quads. **MAP:** 7. **PAGE** 39.

Trail No.	Trail Name	Map Loc.	Distance	Difficulty	Beginning Elev.	Ending Elev.	Ranger District
1776 B-C	**Colorado**	L 3	20 Mi.	Easy	10,880'	11,120'	Columbine

Molas Pass to Graysill Lake

ACCESS #1: Go 5.5 miles south of Silverton (44.5 miles north of Durango) on US 550 and turn west on Little Molas Lake Road, then follow for one mile. The Colorado Trail is west of Molas Lake. **ACCESS #2:** Go 28 miles north of Durango on US 550, then turn west at Purgatory Ski Area on to FS Rd. 578 towards Hermosa Park. Follow gravel road for approximately 16 miles to Graysill Lake. Trail can be found on the east end of lake. 4WD suggested for last 7 miles of this road. **ATTRACTIONS:** This portion of the Colorado Trail takes the user across the old Lime Creek Burn (1879) until it reaches timberline. The trail stays at or above timberline until it descends into the Cascade Creek drainage. This area has few trees which allows for excellent views. The trail has a consistent grade that allows for a pleasant trip without excessive exertion. There are plenty of camp sites along this portion of the Colorado Trail. Several other intersecting trails provide the opportunity to plan "loop" trips. **USE:** Moderate. **ACTIVITIES:** HIKING. **USGS:** Hermosa Peak, Engineer Mountain, Snowdon Peak Quads. **MAPS:** 2,3,6,&7. **PAGES:** 35,36,38&39.

Trail No.	Trail Name	Map Loc.	Distance	Difficulty	Beginning Elev.	Ending Elev.	Ranger District
1776 C-D	**Colorado**	J 3	31 Mi.	More Diff.	11,120'	11,600'	Columbine

Graysill Lake to Kennebec Pass

ACCESS #1: (Graysill Lake) Go 28 miles north of Durango on U.S. 550 and turn west at Purgatory Ski Area onto FDR 578 towards Hermosa Park. The road crosses Hermosa Creek about 8 miles from U.S. 550. Continue for 7 miles to get to Graysill Lake. Trail is on eastside of lake. 4WD is suggested on last 7 miles of this road. **ACCESS #2:** (Hotel Draw) Follow above directions. One mile past the crossing of Hermosa Creek, take left fork onto FDR 550, off FDR 578. Follow FDR 550 about 7 miles. Trail is signed on roadway. 4WD suggested last 8 miles. **ACCESS #3:** (Kennebec Pass) About 11 miles west of Durango on US 160, turn north on FS Rd. 571. Trailhead is about 14 miles from U.S. 160. The last few miles is an unmaintained 4WD road. **ACCESS #4:** FDR 564 crosses the Colorado Trail in a number of places. In addition, the Colorado Trail can be accessed off the end of FDR 436 by following a short 1 mile connecting trail. **ATTRACTIONS:** From Graysill Lake to Hotel Draw, the trail crosses Blackhawk Pass where the views are spectacular and the hike follows the dividing ridge between the Hermosa and Dolores drainages. **NOTE:** Water is limited and watch for bad weather on the ridge. Though this portion of the trail is generally well marked it may be obscure in some places due to intersections with FDR 564 until it starts onto Indian Ridge Trail which is easier to follow. **USE:** Moderate. **ACTIVITIES:** HIKING. **USGS:** Hermosa Peak, Elk Creek, Orphan Butte, La Plata Quads. **MAPS:** 5,6&9. **PAGES:** 37, 38,&40.

Trail No.	Trail Name	Map Loc.	Distance	Difficulty	Beginning Elev.	Ending Elev.	Ranger District
1776 D-E	**Colorado**	J 5	21 Mi.	More Diff.	11,600'	6,960'	Columbine

Kennebec Pass to Junction Creek
ACCESS #1: (Kennebec Pass) Go 11 miles west of Durango on US 160 and turn north on FS Rd. 571. The trailhead is about 14 miles from US 160. The last few miles is an unmaintained 4WD road. **ACCESS #2:** (Junction Creek) Turn west on 25th Street in Durango. Go about 3 miles to the forest boundary and the trail begins off to the left (west). There is more adequate parking 1.1 miles up the road (FS Rd. 171) at the first switchback. Trail is on the west side of the road. **ACCESS #3:** Continue beyond Access #2 on FS Rd. 171. Approximately 17.5 miles from the forest service boundary, there is a side road to the south. Turn and follow for 0.7 miles and the trail crosses the road. **ATTRACTIONS:** The drop in elevation from Kennebec Pass to Junction Creek Trail is 4,790 ft. which is the greatest single altitude change on the Colorado Trail. There may be some confusion on trail location about 1/2 mile below Access #3 due to right-of-way problems across a mine claim. Part of this trail follows the historic Oro Fino (Fine Gold) Trail. Keep in mind two hazards: flash floods can occur in the flood plains and there is poison ivy. Water is scarce until reaching canyon floor. **USE:** Heavy. **ACTIVITIES:** HIKING. **USGS:** La Plata, Monument Hill, Durango East Quad.s. **MAP:** 9&10. **PAGE:** 41.

Trail No.	Trail Name	Map Loc.	Distance	Difficulty	Beginning Elev.	Ending Elev.	Ranger District
1776 LOOP	**Colorado**	L 3	24 Mi.	Moderate	10,900'	9,500'	Columbine

Colorado - South Mineral
ACCESS: Follow U.S. 550 north from Durango to Molas Pass and the turnoff to the Little Molas Lake. Find the trail where it crosses FDR 584. **ATTRACTIONS:** Use of the topographic maps may be important to determine when to leave the Colorado Trail and turn down South Mineral Creek. **USE:** Moderate. **ACTIVITIES:** HIKING MOUNTAIN BIKES. **USGS:** Silverton, Ophir, Snowdon Peak, Engineer Mountain Quads. **MAPS:** 2,3,6&7. **PAGES:** 35, 36, 38&39.

METHOD FOR RATING TRAIL DIFFICULTY

Four categories for degree of difficulty are as follows:

EASY:
A.   Route is most level with short uphill/downhill sections.
B.   Excellent to good tread surface and clearance.
C.   Absence of navigational difficulties/hazards.

MODERATE:
A.   Route is level to sloping with longer uphill/downhill sections.
B.   Good-to-fair surface and clearance.
C.   Minimal navigational difficulties/hazards.

MORE DIFFICULT:
A.   Route is level to steep with sustained uphill/downhill sections.
B.   Fair to poor surface and clearance.
C.   Short sections involving significant navigational difficulties/hazards.

MOST DIFFICULT:
A.   Route is mostly steep with sustained uphill/downhill sections.
B.   Poor-to-nonexistent tread surface and clearance.
C.   Longer sections involving significant navigational difficulties/hazards.

Any rating (i.e., Moderate trail difficulty) found in this book is based on the above scale which has been established for Forest Service purposes.

NOTE: Information contained in this guide is for general recreation reference and trip planning only. Use good judgement! When planning a trip physical condition, age, altitude and weather conditions should be considered. Outdoor Books is not responsible for any injury or mishap resulting in the use of this guide for other than its intended use.

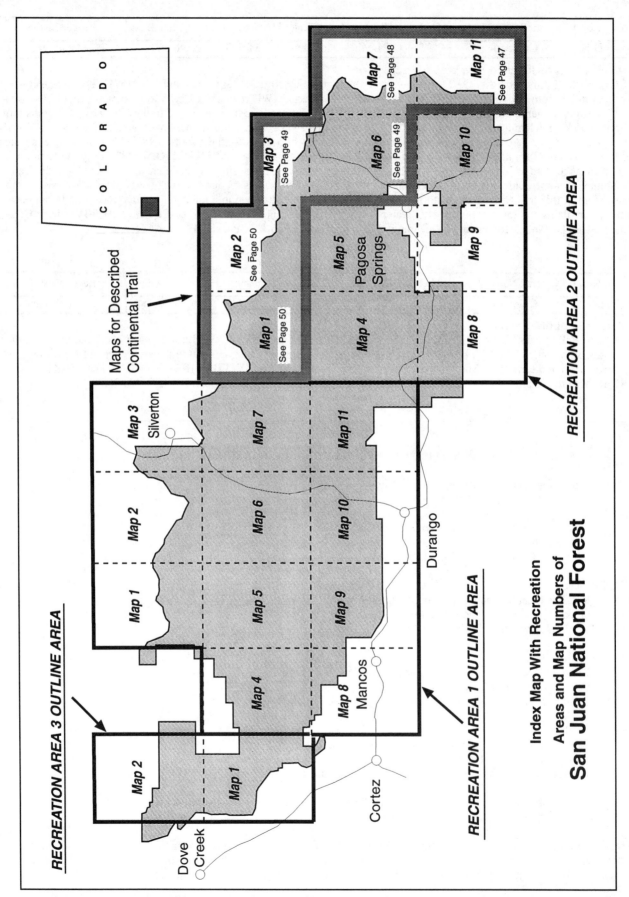

Index Map With Recreation
Areas and Map Numbers of
San Juan National Forest

Ties to Map 7 Page 48

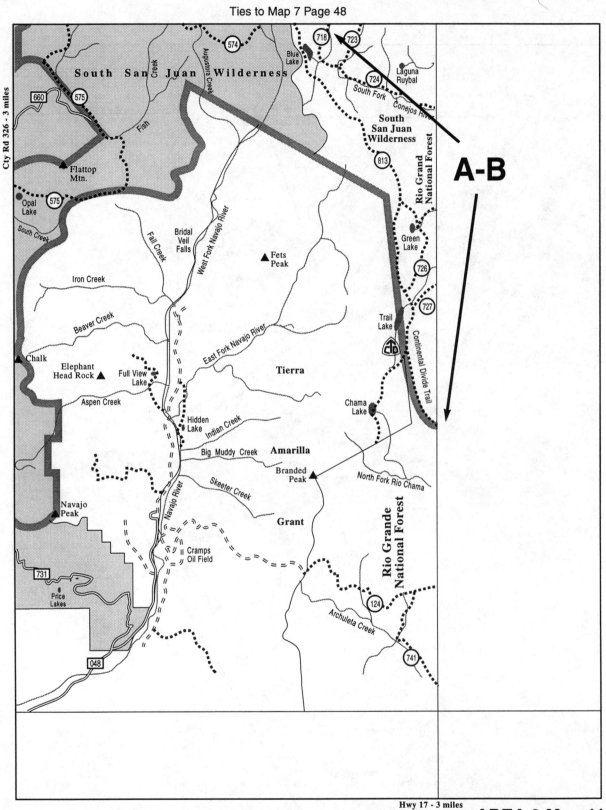

South San Juan Wilderness

Cty Rd 326 - 3 miles

660
575

South San Juan Wilderness

574
Blue Lake
718
723
724
Laguna Ruybal
South Fork
Conejos River

Rio Grand National Forest

813

A-B

Flattop Mtn.

Opal Lake
575

South Creek

Fish Creek
Augustina Creek

Bridal Veil Falls

Fall Creek

West Fork Navajo River

Fets Peak

Green Lake
726

727

Iron Creek

Beaver Creek

East Fork Navajo River

Trail Lake
CDT
Continental Divide Trail

Chalk
Elephant Head Rock
Full View Lake
Aspen Creek

Tierra

Chama Lake

Hidden Lake
Indian Creek

Big Muddy Creek

Amarilla

Branded Peak
North Fork Rio Chama

Navajo Peak

Skeeter Creek
Navajo River

Grant

Cramps Oil Field

Rio Grande National Forest

731

Price Lakes

124

Archuleta Creek

048

741

Hwy 17 - 3 miles

AREA 2 Map 11

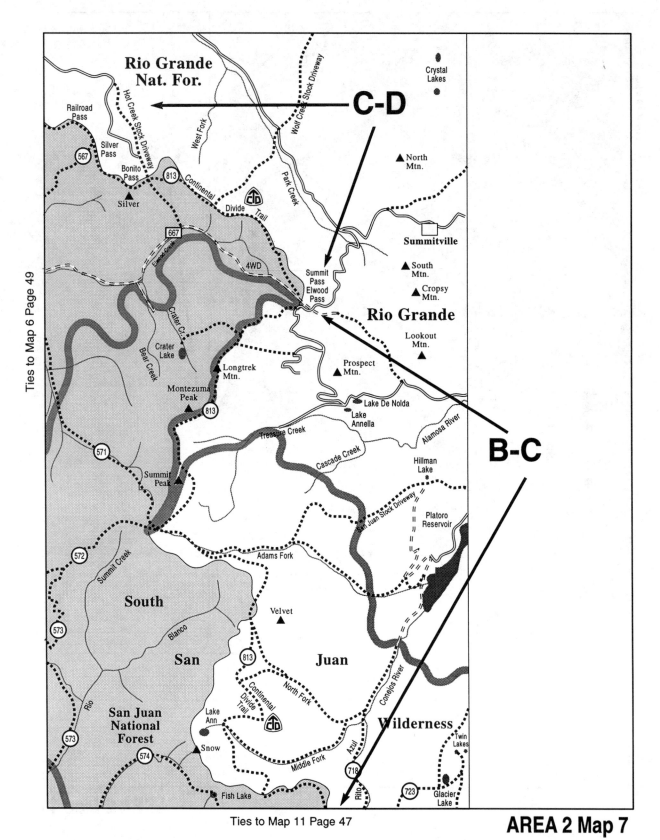

Rio Grande
Nat. For.

Crystal
Lakes

Railroad
Pass

C-D

Silver
Pass

567

North
Mtn.

West Fork

Wolf Creek Stock Driveway

Hot Creek Stock Driveway

Bonito
Pass

813

Continental

Park Creek

Summitville

Silver

Divide

Trail

667

4WD

Summit
Pass
Elwood
Pass

South
Mtn.

Cropsy
Mtn.

Rio Grande

Elwood Creek

Ties to Map 6 Page 49

Crater Cr.

Bear Creek

Crater
Lake

Longtrek
Mtn.

Lookout
Mtn.

Montezuma
Peak

813

Prospect
Mtn.

Lake De Nolda

Lake
Annella

571

Treasure Creek

Alamosa River

B-C

Cascade Creek

Hillman
Lake

Summit
Peak

San Juan Stock Driveway

Platoro
Reservoir

572

Summit Creek

Adams Fork

South

Velvet

573

Blanco

San

813

Juan

North Fork

Conejos River

San Juan
National
Forest

Rio

Lake
Ann

Continental
Divide
Trail

Wilderness

573

Snow

Twin
Lakes

574

Middle Fork

Azul

718

Rito

723

Glacier
Lake

Fish Lake

Ties to Map 11 Page 47

AREA 2 Map 7

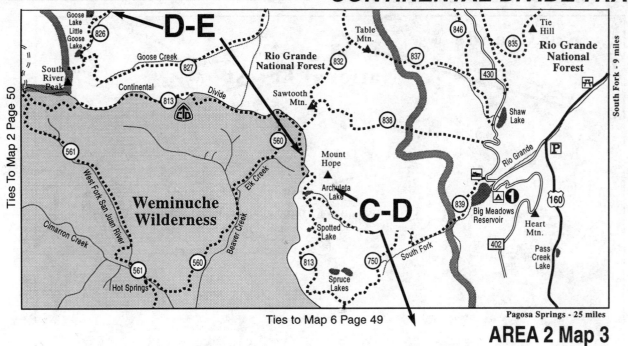

AREA 2 Map 3

Goose Lake
Little Goose Lake

826

Goose Creek

827

South River Peak

Continental   Divide

813

561

Weminuche Wilderness

West Fork San Juan River

Cimarron Creek

Beaver Creek

Elk Creek

560

561

560

Hot Springs

D-E

Rio Grande National Forest

Table Mtn.

832

837

846

835

Tie Hill

430

Rio Grande National Forest

Sawtooth Mtn.

838

Shaw Lake

Mount Hope

Archuleta Lake

C-D

Spotted Lake

813

750

Spruce Lakes

South Fork

839

Big Meadows Reservoir

①

402

Heart Mtn.

160

Pass Creek Lake

Rio Grande

P

South Fork - 9 miles

Ties To Map 2 Page 50

Ties to Map 6 Page 49

Pagosa Springs - 25 miles

AREA 2 Map 6

Ties to Map 3 Page 49

561

West Fork

Rock Lakes

Sheep Mtn.

Lake Creek

Navajo Trail

Continental

Divide Trail

Del Norte - 32 miles

Tucker Ponds

④

Rio Grande National Forest

Weminuche Wilderness

580

Saddle Mtn.

Hatcher Lakes

①

②

160

Wolf Creek

563

Treasure Falls

Windy Pass Trail

039

725

Treasure Mtn.

Wolf Creek Pass

Wolf Creek

Treasure Pass

Alberta Peak

813

Albe Park Reservoir

C-D

Treasure Mountain Trail

Lane Creek

565

567

570

East Fork

160

Turkey Creek

037

Snowball Creek

4WD

667

③

Turner

570

581

Sand Creek

Peer Creek

Johnny Creek

East Fork San Juan River

570

Ties to Map 5 Page 59

Tie to Map 7 Page 48

CONTINENTAL DIVIDE TRAIL

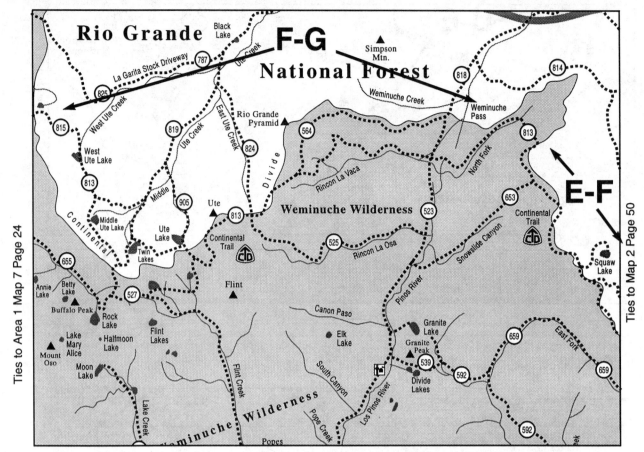

AREA 2 Map 1

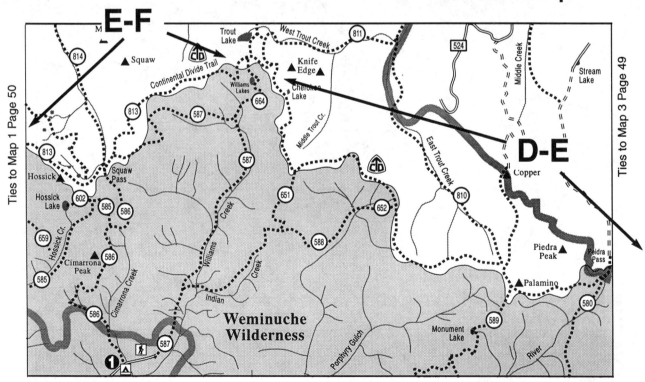

AREA 2 Map 2

Trail No.	Trail Name	Map Loc.	Distance	Difficulty	Beginning Elev.	Ending Elev.	Ranger District
813 A-B	**Continental Divide**	V 7	19 Mi.	Easy/Diff.	10,600'	11,450'	Conjes Peak*

Cumbres Pass to Blue Lake   (* Rio Grand Nat. Forest)
ACCESS: Follow Highway 17 to the summit of Cumbres Pass. Immediately north of the summit, turn northwest onto FDR 119 which passes behind the old railroad station and by a small pond. Continue on FDR 119 to its end, approximately 3 miles. The Continental Trail begins at the roads' end. Through an agreement with the landowner, hikers are allowed to cross the one mile of private land at the beginning of the trail. **ATTRACTIONS:** FDT 813 is a section of the Continental Divide Trail which continues north and south through the Rocky Mountains. In addition to traversing a unique terrain, the trail intersects with a number of other trails, allowing one to make side trips into the many drainages dropping from the divide. From high on the Continental Divide, views of the alpine plateau below, the many tributary drainages of the Conejos River, the San Luis Valley, and Sangre de Cristos, far in the distance can be seen. After reaching the east of the divide, a beautiful view of the Chama basin can be seen. This view is evident for several miles. As Trail Lake is approached, a view of the main canyon of the Navajo demands attention to the west. At the top of Flat Mountain, one enters the South San Juan Wilderness. Beyond this point motorized equipment is prohibited. The Continental Divide Trail remains at high elevations for its entire length. Snow conditions in early summer and early fall may make passage difficult. Hikers should also be prepared for harsh weather conditions that may be encountered at anytime at these elevations. **NARRATIVE:** FDT 813 is but a short segment of the entire Continental Divide Trail but it offers a unique experience to hikers because the terrain differs considerably from that of most of the divide. In the southern San Juan Mountains, the Continental Divide follows the peaks which break the gently rolling alpine plateau to the east. The plateau is dotted with numerous lakes and cut by major tributaries of the Conejos and Navajo Rivers such as Elk Creek South Fork of the Conejos, and Canon Verde. These drainages provide not only variety to the terrain and a change in ecosystems but also interesting side trips if one hikes into them from the Divide Trail. Some of the districts' best fishing is found in the lakes along the Continental Divide Trail and the streams draining them. **USE:** Use and quality of the trail vary considerably. Within the Conejos District, the trail is generally well used and well marked. From the Continental Divide Trail, one can also join the Fish Lake Trail that descends the North Fork of Fish Creek to the west, within the San Juan National Forest. **ACTIVITIES:** HIKING. **USGS:** Victoria Lake, Elephant Head Rock Quads. **MAP:** Area 2, Map 11, Page 47.

Trail No.	Trail Name	Map Loc.	Distance	Difficulty	Beginning Elev.	Ending Elev.	Ranger District
813 B-C	**Continental Divide**	U 6	22.9 Mi.	Easy/Diff.	11,600''	11,450'	Conjes Peak*

Elwood Pass to Blue Lake   (* Rio Grand Nat. Forest)
ACCESS:ACCESS #1: Travel 23 miles (37 km) west on Highway 160 from Del Norte Ranger Station (7 miles (11 km) west of South Fork) to Park Creek access FDR 380. Turn left and follow Park Creek Road for approximately 17 miles (27 km) to the old Elwood Guard Station (cabin). The trail is about 1/4 mile (.4 km) west of the cabin. **ACCESS #2:** Wolf Creek Pass on U.S.160 20 miles (32 km) southwest of South Fork. **ATTRACTIONS:** The Continental Divide Trail is part of the Continental Divide National Scenic Trail which will start at the Canadian Border on the north and continue to the Mexican Border on the south. The portion of this trail which passes through the Conejos Peak Ranger District is almost entirely within the boundaries of the South San Juan Wilderness area. The visitor should become familiar with Wilderness regulations before entering. The trail at present is poorly maintained and lacks adequate signing in places. Because of it's condition, it is suggested only experienced hikers attempt this trip. Horseback riders should he very cautious at the present time due to the numerous potentially dangerous areas. Visitors who do use the trail will be rewarded for efforts by spectacular vistas in all directions. Due to the exposed nature of most of this trail, extremely close watch should be kept on weather conditions. **NARRATIVE:** The Continental Divide Trail continues onto the Conejos Peak Ranger District, just off FDR 380 at Elwood Pass. The trail continues to the Conejos Ranger District boundary at Blue Lake on the south. The trail is poorly maintained and inadequately signed most of the way. There are stretches of the trail which pose dangers to the careless or inexperienced hiker. Horse travel is dangerous at present due to these conditions which pose dangers even to riderless horses. Suitable camping areas are not numerous but an adequate number do exist. Water is available along the trail, but should be treated prior to use. The trail is above timberline most of the way. Sudden and violent storms can he expected. The visitor should be prepared for these and always watchful for their approach. The trail offers access to Crater Lake, Lake Arm, and Blue Lake. Numerous trails provide access to the Continental Divide Trail. These offer numerous loop possibilities for trips of varying lengths. For additional trails which offer access to the Continental Divide Trail, check the Del Norte and Conejos Ranger Districts of the Rio Grande National Forest, and the San Juan National Forest. **USE:** Medium. **ACTIVITIES:** HIKING, HORSES. **USGS:** Elwood Pass, Summut Peak Quads. **MAP:** Area 2, Map 7, Page 48.

CONTINENTAL DIVIDE TRAIL

Trail No.	Trail Name	Map Loc.	Distance	Difficulty	Beginning Elev.	Ending Elev.	Ranger District
813 C-D	**Continental Divide**	U 5	25 Mi.	Mod/Diff.	11,630"	12,400'	Del Norte*

Elwood Pass to Sawtooth Mountain   (* Rio Grand Nat. Forest)

ACCESS #1: Travel 23 miles (37 km) west on U.S. 160 from Del Norte Ranger Station (7 miles (11 km) west of South Fork) to Park Creek access road (FDR 380). Turn left and follow Park Creek Road for approximately 17 miles (27 km) to the old Elwood Guard Station (cabin). The trail is about 1/4 mile (.4 km) west of the cabin. **ACCESS #2:** Wolf Creek Pass on U.S.160 20 miles (32 km) southwest of South Fork. **ATTRACTIONS:** This is a small portion of the Continental Divide Trail which runs from Canada to Mexico. Much of this portion of the trail is at or above timberline. Many years find snow present until late June or July and severe weather can occur at any time. The high elevation and rugged terrain make this a moderate to difficult trail. The scenery along the trail is spectacular. **NARRATIVE:** As this trail winds along the Continental Divide from Canada to Mexico, it passes through a wide variety of topography and life zones. This particular portion of the trail, from Elwood Pass to Sawtooth Mountain, crosses some of the highest elevations on the trail. The trail is in good condition and not extremely steep. It is open to both foot and horse travel. In 1989, portions of this trail were relocated and reconstructed for user convenience and safety. **USE:** Moderate. **ACTIVITIES:** HIKING, HORSES. **USGS:** Elwood Pass, Wolf Creek Pass, Spar City Quads Quads. **MAPS:** Area 2, Maps 3, 6&7, Pages 48&49.

Trail No.	Trail Name	Map Loc.	Distance	Difficulty	Beginning Elev.	Ending Elev.	Ranger District
813 D-E	**Continental Divide**	S 4	21.3 Mi.	Mod/Diff.	11,630"	12,400'	Del Norte*

Sawtooth Mountain to Knife Edge   (* Rio Grand Nat. Forest)

ACCESS: Travel southwest from Creede on Colorado Highway 149 for 7 miles to the junction of Colorado 149 and FDR 523 (Middle Creek Road), then 9 miles on this road to the Trout Creek Trailhead. Proceed on the Main Trout Creek and West Trout Creek Trail to Trout Lake. The Continental Divide Trail (FDT 813) can be seen south and above Trout Lake as it traverses the fabled Knife Edge. **ATTRACTIONS:** Beautiful views, isolated wild country, elevations approaching 13,000 feet, and spine-tingling climbs along rocky cliffs await the hiker of this trail segment. The hiker who likes to fish will find only one fishable lake within two miles of the trail. Goose Lake not only is good fishing, but has several nice camping areas. Little Goose Lake is a dead lake. Good camping areas, with comfortable tree cover are fairly frequent between the Knife Edge and Piedra pass. From the Pass to Sawtooth Mountain, one may have to wander away from the trail a short distance to find a nice flat camping area with tree cover and ready water supply. Hikers usually have this segment of the Continental Divide Trail all to themselves, as not too many people visit this part of the Continental Divide during the summer months. Several trail junctions are not well-defined or signed, so be sure to have a U.S.G.S. topographic map to help guide yourself. If you find yourself wandering more than a mile away from the Continental Divide, stop to check your map to make sure you did not start down one of the ten trails that junction with this segment of the Continental Divide Trail. **NARRATIVE:** This trail segment begins with a 1/4 mile traverse of the cliffs that form the fabled Knife Edge. Crossing the Knife Edge is not as scary as it used to be, as the narrow, rocky trail tread of the past has been "improved" to reduce hazardous conditions. From the Knife Edge to Piedra Pass, the trail will go up and down like a roller coaster. Most of the "ups' and "downs" are reasonably short steep pitches. The trail slopes quite close to the Continental Divide and frequently weaves above and below timberline. From the water diversion ditches at Piedra Pass, the trail will make a rather lengthy long climb towards South River Peak. From the South River Peak Area to Sawtooth Mountain, the trail will wind along, the crest of the Continental Divide, well above timberline. Some of the highest elevations along the trail occur in this segment of trail. Following this section of trail is occasionally difficult in places as some sections of trail tread can be indistinct. **USE:** Moderate to light. **ACTIVITIES:** HIKING, HORSES. **USGS:** Cimarrona Peak, Palimino Mountain, South River Quads. **MAPS:** Area 2, Maps 2&3. Pages 49&50.

Trail No.	Trail Name	Map Loc.	Distance	Difficulty	Beginning Elev.	Ending Elev.	Ranger District
813 E-F	**Continental Divide**	Q 3	17.2 Mi.	Mod/Diff.	10,600'	11,800'	Creede*

Knife Edge to Weminuche Pass   (* Rio Grand Nat. Forest)
ACCESS: Travel southwest from Creede on Hwy. 149 for approximately 20.1 miles to the junction of Hwy. 149 and FDR 520 (Upper Rio Grande River Road), then approximately 10 miles to Thirty Mile Campground. Take the Weminuche Creek Trail (FDT 818) from the campground to Weminuche Pass. Once at the Pass, cross the diversion ditch and stay to the east side of the headwaters of the Pine River and you should have little trouble picking up the Continental Divide Trail (FDT 813). **ATTRACTIONS:** Beautiful panoramic views exist along this segment of the Continental Divide Trail. Fishermen will find some short side trips can be made to Trout, Williams, and Squaw Lakes. A side trip to Squaw Lake will provide a good camping area and good fishing, but a steep climb back to the Continental Divide Trail. Side trips to either Trout or Williams Lakes do not involve hard climbs back to the Continental Divide Trail. Much of this trail segment is above timberline. Good camping areas, with tree cover can be found at Weminuche Pass, the head of the North Fork of the Pine, and at Squaw Pass. Bighorn Sheep can occasionally be observed in the area near Hossick Peak. Elk can also be observed grazing at times in many of the meadow areas. The best areas for seeing elk are usually near Chief Mountain and at the headwaters of Little Squaw Creek. Elevations are high and the air is thin, so plan to be in good physical condition. It is usually best to do your hiking in the early morning to early afternoon hours. Afternoon thundershowers can be severe, with few areas along the trail providing protection from lightning hazards. **NARRATIVE:** Shortly after leaving the tree cover at Weminuche Pass, the trail will cross a water diversion ditch and proceed through a wet, boggy meadow for about one mile. The trail will then turn easterly and proceed up the North Fork of the Pine River. It will climb gradually through the spruce-covered slopes of this drainage before getting above timberline and crossing the boggy meadows and willow fields at the head of Snowslide Canyon. The trail will continue along a broad open grassy ridge along the Continental Divide for several miles. It will then begin making a gradual descent through open meadows and parks below the rugged rocky cliff-like country near Hossick Peak. After Squaw Pass, the trail will make a rather strong climb through scattered spruce timber patches before entering the open grassy ridges near Chief Mountain. The trail will continue along the backbone of the Continental Divide, weaving back and forth from one side to the other until it reaches the open pass between Williams Lake and Trout Lake. The Knife Edge lies directly ahead, with the trail carved into the side of a cliff, jutting sharply out from the Continental Divide. The trail is well defined for the vast majority of this segment. Some portions of this trail segment are poorly located on the Forest Service 1/2' mile map. **USE:** Moderate to Heavy. **ACTIVITIES:** HIKING, HORSES. **USGS:** Cimarrona Peak, Granite Lake, Little Squaw Creek, Weminuche Pass.   **MAPS:** Area 2, Maps 2&3Pages 49&50.

Trail No.	Trail Name	Map Loc.	Distance	Difficulty	Beginning Elev.	Ending Elev.	Ranger District
813 F-G	**Continental Divide**	P 3	20.4 Mi.	Mod/Diff.	12,500'	10,600'	Creede*

Weminuche Pass to Hunchback Pass   (* Rio Grand Nat. Forest)
ACCESS: Travel southwest from Creede on Hwy 149 for approximately 20.1 miles to the junction of Hwy149 with FDR 520 (Upper Rio Grande River Road), then approximately 26 miles to the Beartown Road (FDR 506) junction. Proceed up the Beartown Road to 1/4 mile below Kite Lake, where the Continental Divide Trail (FDT 813) crosses the road. Hike 3/4 mile southwest to Hunchback Pass. A 4 WD vehicle is required for access to the Kite Lake Area. **ATTRACTIONS:** The opportunity to see and visit the towering Rio Grande Pyramid Mountain peak and the adjacent "Window" area in this trail segment. The extremely rugged, ragged beauty of the Needles Mountain area of the Weminuche Wilderness can also be viewed in the distance from the trail, when traveling from Hunchback Pass to West Ute Lake. Access to many fishable lakes in the Upper Vallecito and Ute Creek Drainages is provided by this trail segment. West Ute, Middle Ute, Twin Ute Lakes, Ute Lake, Rock Lake, and Flint Lakes either adjoin the trail or are a quick side trip. People pressures are high from mid-July to mid-August along the Continental Divide at the head of the Vallecito and Ute Creek Drainages. This trail segment is well above timberline for all but a one or two mile stretch near Weminuche Pass. The best camping areas are off the trail and at or below timberline where trees provide protection, comfort, variety and fuel wood. **NARRATIVE:** The trail from Hunchback Pass begins by making a fairly steep 1 mile descent into the Vallecito Drainage before making a steady climb up Nebo Creek. Approximately 1/2 mile up Nebo Creek, the trail will turn and pass through a flat open meadow below the towering ragged cliffs of Mount Nebo. The trail will continue through grassy open meadows and slopes before reaching West Ute Lake. The trail will then switchback up the rocky slopes south of the Lake before reaching the divide between West Ute and Middle Ute Drainages. The Trail is very poorly defined from this divide to Twin Ute Lakes. Heavy willow growth and wet boggy areas make travel difficult to Middle Ute Lake. Most people follow the defined tread of the West Ute Cut-Off Trail to the Main Ute Trail below Twin Ute Lakes. They rejoin the Continental Divide Trail to the Main Ute Trail below Twin Ute Lakes. They rejoin the Continental Divide Trail at Twin Ute Lakes and proceed over hilly open country toward Main Ute Lake. After reaching the ridge above Main Ute Lake, the trail will shortly make a steep winding ascent to the headwaters of the Rincon La Osa Drainage. The trail will again become poorly defined at the headwater of Rincon La Osa. Be careful not to go down the well defined Rincon La Osa Trail unless you are looking for a nice camping area in the tree lined meadows. By staying close to the Continental Divide, you will eventually pick up the well defined trail tread heading up the steep slope below the "Window." The trial will cross the Divide below the "Window." Once again, hikers need to be careful not to go down the Rincon La Vaca Trail but should bear northward across the open meadows below the "Window" and "The Rio Grande Pyramid." Shortly, the well defined tread of the High Line Trail will come into view. Following this trail will take one along the narrow cliffs before dropping sharply to Weminuche Pass. **USE:** Mod. to heavy. **ACTIVITIES:** HIKING, HORSES. **USGS:** Rio Grande, Weminuche Pass Quads. **MAPS:** Area 2, Maps1&2, Page 50.

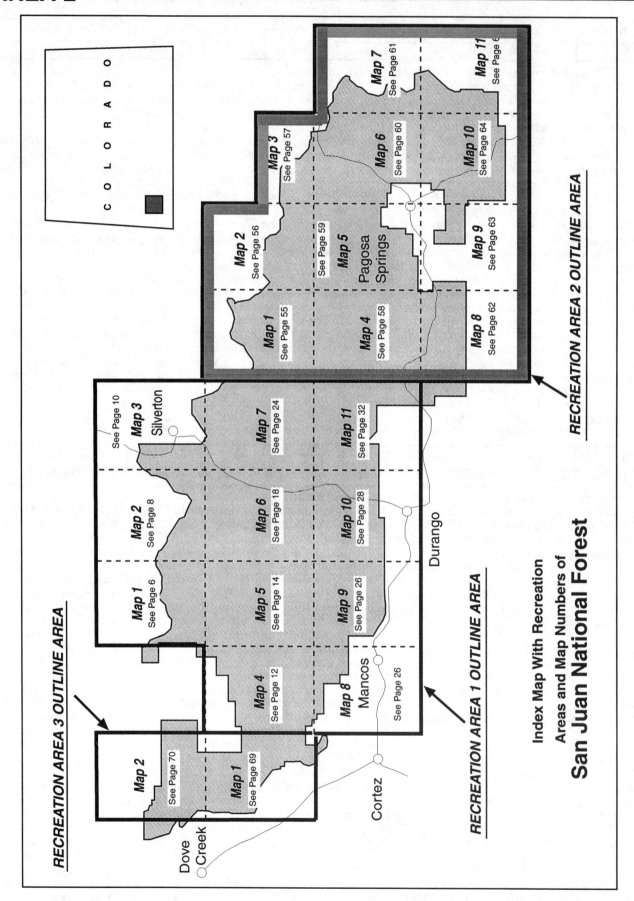

RECREATION AREA 2 OUTLINE AREA

Map 7
See Page 61

Map 11
See Page 6

Map 3
See Page 57

Map 6
See Page 60

Map 10
See Page 64

Map 2
See Page 56

See Page 59

Map 5
Pagosa Springs

Map 9
See Page 63

Map 1
See Page 55

Map 4
See Page 58

Map 8
See Page 62

COLORADO

RECREATION AREA 3 OUTLINE AREA

See Page 10

Map 3
Silverton

Map 7
See Page 24

Map 11
See Page 32

Map 2
See Page 8

Map 6
See Page 18

Map 10
See Page 28

Map 1
See Page 6

Map 5
See Page 14

Map 9
See Page 26

Durango

Map 4
See Page 12

Map 8
Mancos
See Page 26

Map 2
See Page 70

Map 1
See Page 69

Cortez

Dove
Creek

RECREATION AREA 1 OUTLINE AREA

Index Map With Recreation
Areas and Map Numbers of
San Juan National Forest

NO CAMPGROUNDS LOCATED IN AREA 2 MAP 1.

NO TRAIL NARRATIVE THIS MAP

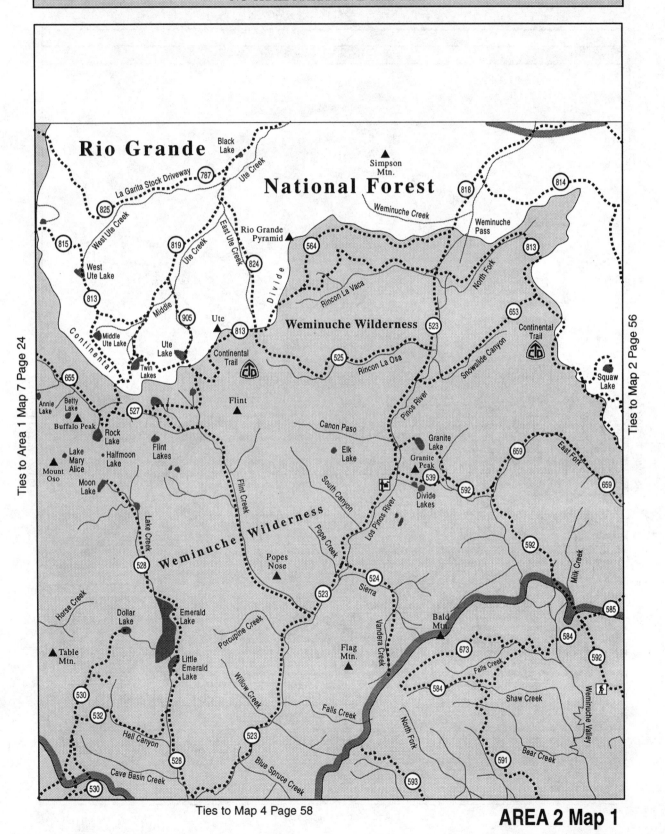

Ties to Area 1 Map 7 Page 24

Ties to Map 2 Page 56

Ties to Map 4 Page 58

AREA 2 Map 1

AREA 2

Map No.	Name	Fee	No. of Units	Max. Length	Elev.	Toilets	Water	Ranger District
CAMPGROUNDS LOCATED IN AREA 2 MAP 2.								
1.	Cimarrona	$	21	35'	8,400'	Yes	Yes	Pagosa

TRAIL NARRATIVE #587 PAGE 66
</>

Little Ruby Lake

Fuchs Reservoir

Ruby Lake

Baldy Lake

Ruby

Little Sqauw

Driveway

815

816

889

811

525

Love Lake

Fern

Creek

Stock

Weminuche Wilderness

Red Lakes

816

Rio Grande National Forest

Trout Creek

Chief Mtn.

814

Squaw

Trout Lake

West Trout Creek

811

524

Stream Lake

Continental Divide Trail

Knife Edge

Williams Lakes

813

587

664

Cherokee Lake

Middle Trout Cr.

Middle Creek

587

813

Squaw Pass

Hossick

Hossick Lake

602

585

586

651

652

Copper

659

Hossick Cr.

586

Williams

Creek

588

810

Piedra Peak

Peidra Pass

585

Cimarrona Peak

Indian

Creek

Weminuche Wilderness

Palamino

586

Porphyry Gulch

Monument Lake

589

580

587

1

River

640

Window Lake

590

Puerto Blanca Creek

Teal R.A.

2

Williams Creek Reservoir

Palisade Lakes

Lean Creek

589

Sugarloaf Mtn.

Red Mtn

583

638

636

Toner Mtn.

Piedra

Deadman

Ties to Map 1 Page 55

Ties to Map 3 Page 57

Ties to Map 5 Page 59

NO TRAIL NARRATIVE THIS MAP							
CAMPGROUNDS LOCATED IN AREA 2 MAP 3.							
Map No.	Name	Fee	No. of Units	Max. Length	Elev.	Toilets	Water	Ranger District
1.	Big Meadows	$	53	35'	9,500'	Yes	Yes	Divide/Del Norte

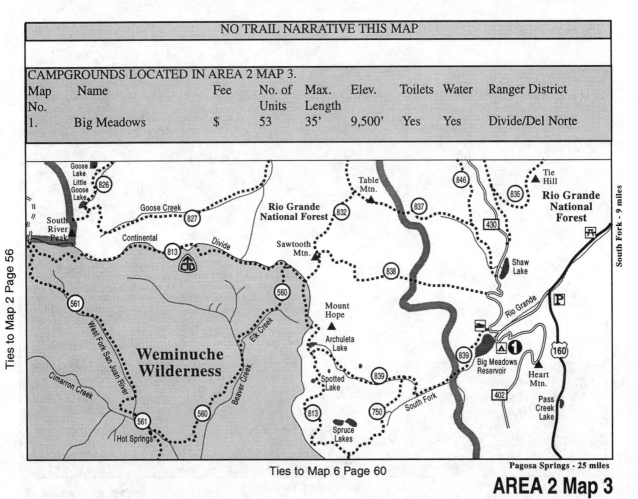

Ties to Map 2 Page 56

Ties to Map 6 Page 60

South Fork - 9 miles

Pagosa Springs - 25 miles

AREA 2 Map 3

AREA 2

Ties to Map 1 Page 55

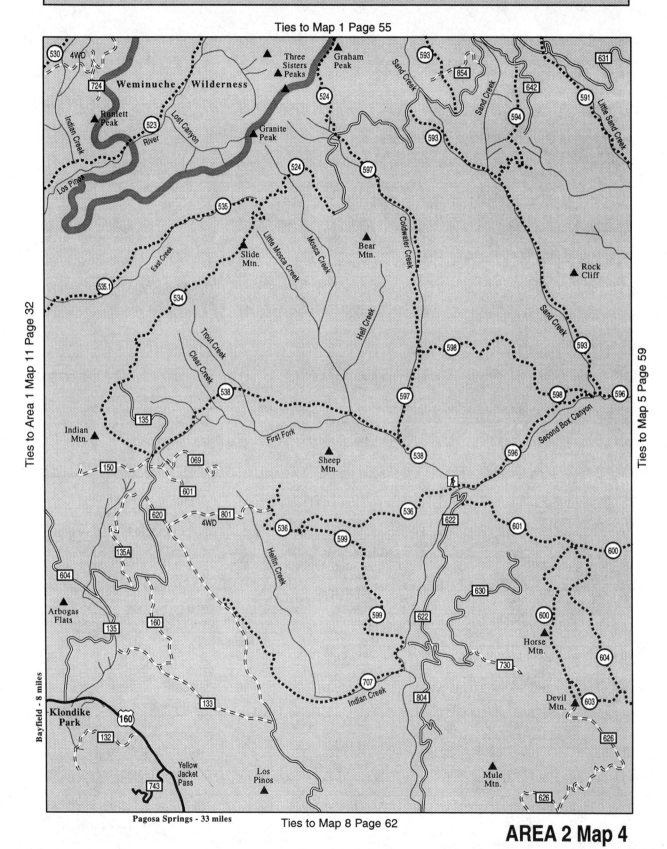

Ties to Area 1 Map 11 Page 32

Ties to Map 5 Page 59

Bayfield - 8 miles

Pagosa Springs - 33 miles

Ties to Map 8 Page 62

AREA 2 Map 4

Map No.	Name	Fee	No. of Units	Max. Length	Elev.	Toilets	Water	Ranger Dist.
1.	Williams Creek	$	66	30'	8,300'	Yes	Yes	Pagosa
2.	Bridge	$	19	50'	7,800'	Yes	Yes	Pagosa

CAMPGROUNDS LOCATED IN AREA 2 MAP 5.

TRAIL NARRATIVE #579 PAGE 66

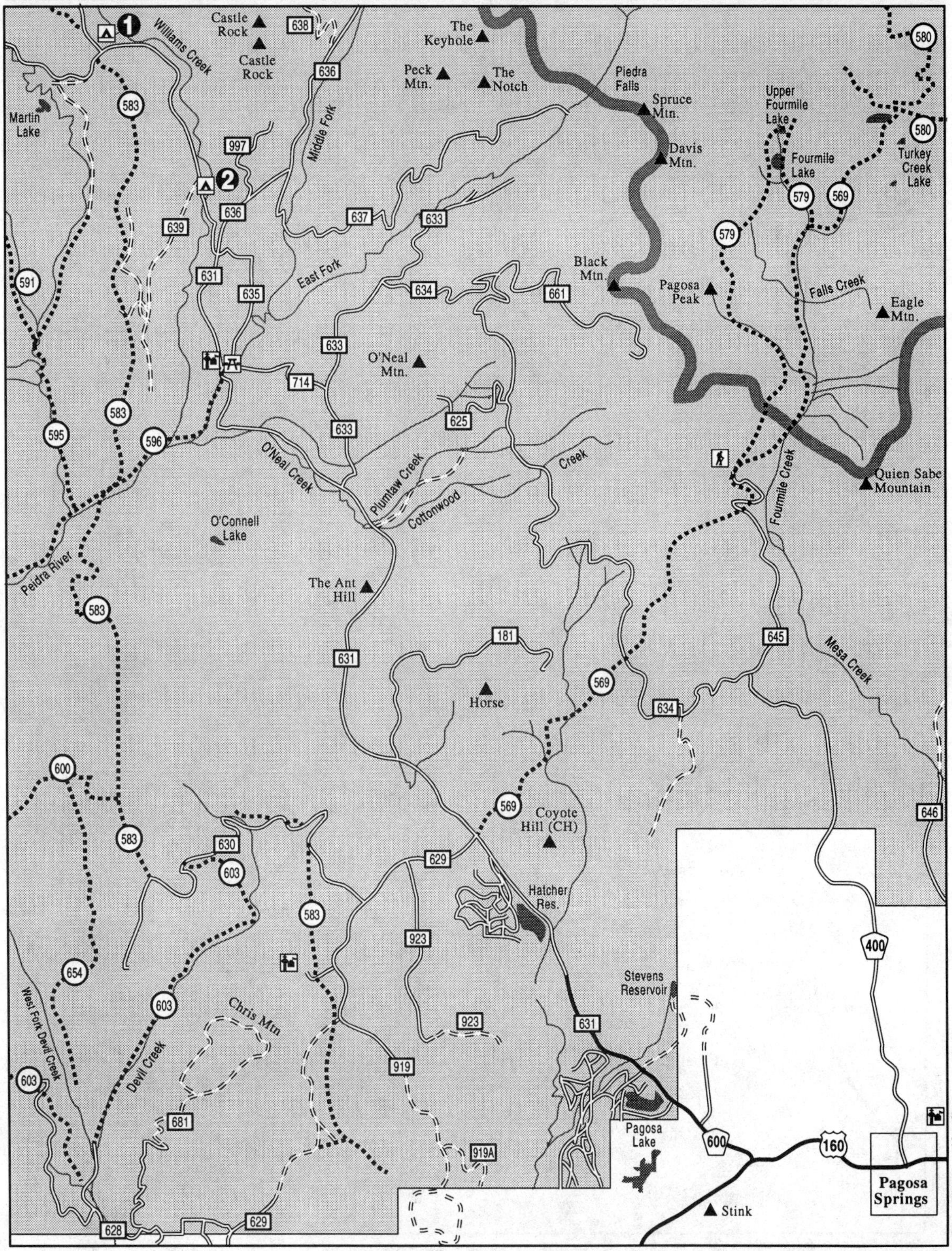

Ties to Map 4 Page 58
Ties to Map 6 Page 60
Ties to Map 9 Page 63

South Fork - 43 miles

AREA 2 Map 5

Map No.	Name	Fee	No. of Units	Max. Length	Elev.	Toilets	Water	Ranger District
1.	West Fork	$	28	35'	8,000'	Yes	Yes	Pagosa
2.	Wolf Creek	$	26	35'	8,000'	Yes	Yes	Pagosa
3.	East Fork	$	26	35'	7,600'	Yes	Yes	Pagosa
4.	Tucker Ponds	$	16	35'	9,600'	Yes	No	Divide/Del Norte

CAMPGROUNDS LOCATED IN AREA 2 MAP 6.

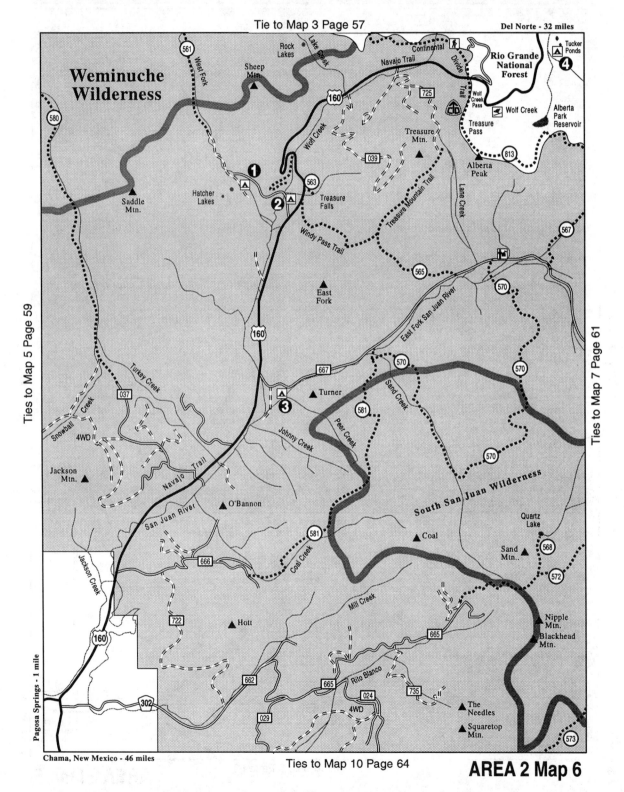

Tie to Map 3 Page 57

Del Norte - 32 miles

Ties to Map 5 Page 59

Ties to Map 7 Page 61

Pagosa Springs - 1 mile

Chama, New Mexico - 46 miles

Ties to Map 10 Page 64

AREA 2 Map 6

NO CAMPGROUNDS LOCATED IN AREA 2 MAP 7

TRAIL NARRATIVE #572 MAPS 6 & 7 PAGE 66

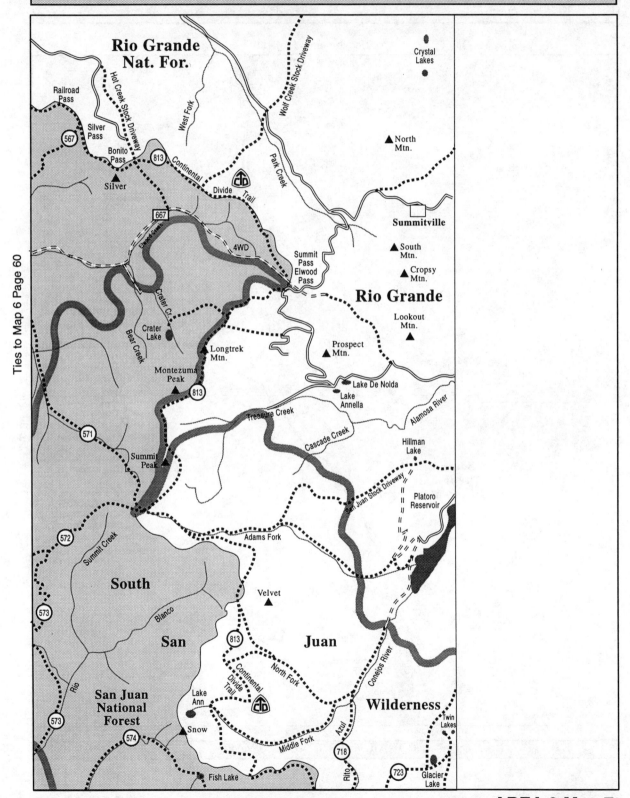

Ties to Map 6 Page 60

Ties to Map 11 Page 65

AREA 2 Map 7

CAMPGROUNDS LOCATED IN AREA 2 MAP 8.							
Map No.	Name	Fee	No. of Units	Max. Length	Elev.	Toilets	Water	Ranger District
1.	Lower Piedra	$	17	35'	7,200'	Yes	No	Pagosa

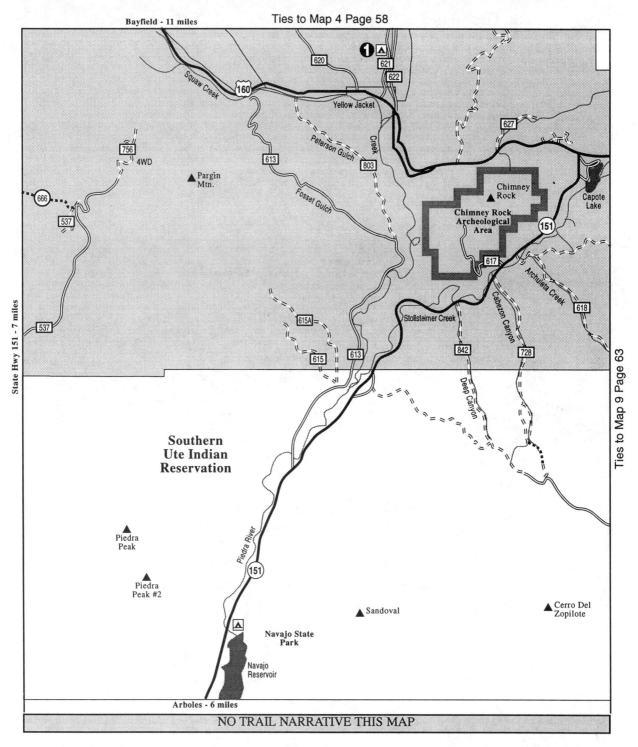

Bayfield - 11 miles   Ties to Map 4 Page 58

Squaw Creek

620
621
622

160

Yellow Jacket

627

756
4WD

613

Peterson Gulch

Creek

803

666

Pargin Mtn.

Chimney Rock

Chimney Rock Archeological Area

Capote Lake

537

Fosset Gulch

151

State Hwy 151 - 7 miles

617

Archuleta Creek

537

615A

Stollsteimer Creek

Cabezon Canyon

618

615
613

842

728

Ties to Map 9 Page 63

Deep Canyon

Southern Ute Indian Reservation

Piedra Peak

Piedra River

Piedra Peak #2

151

Sandoval

Cerro Del Zopilote

Navajo State Park

Navajo Reservoir

Arboles - 6 miles

NO TRAIL NARRATIVE THIS MAP

AREA 2 Map 8

NO CAMPGROUNDS LOCATED IN AREA 2 MAP 9.

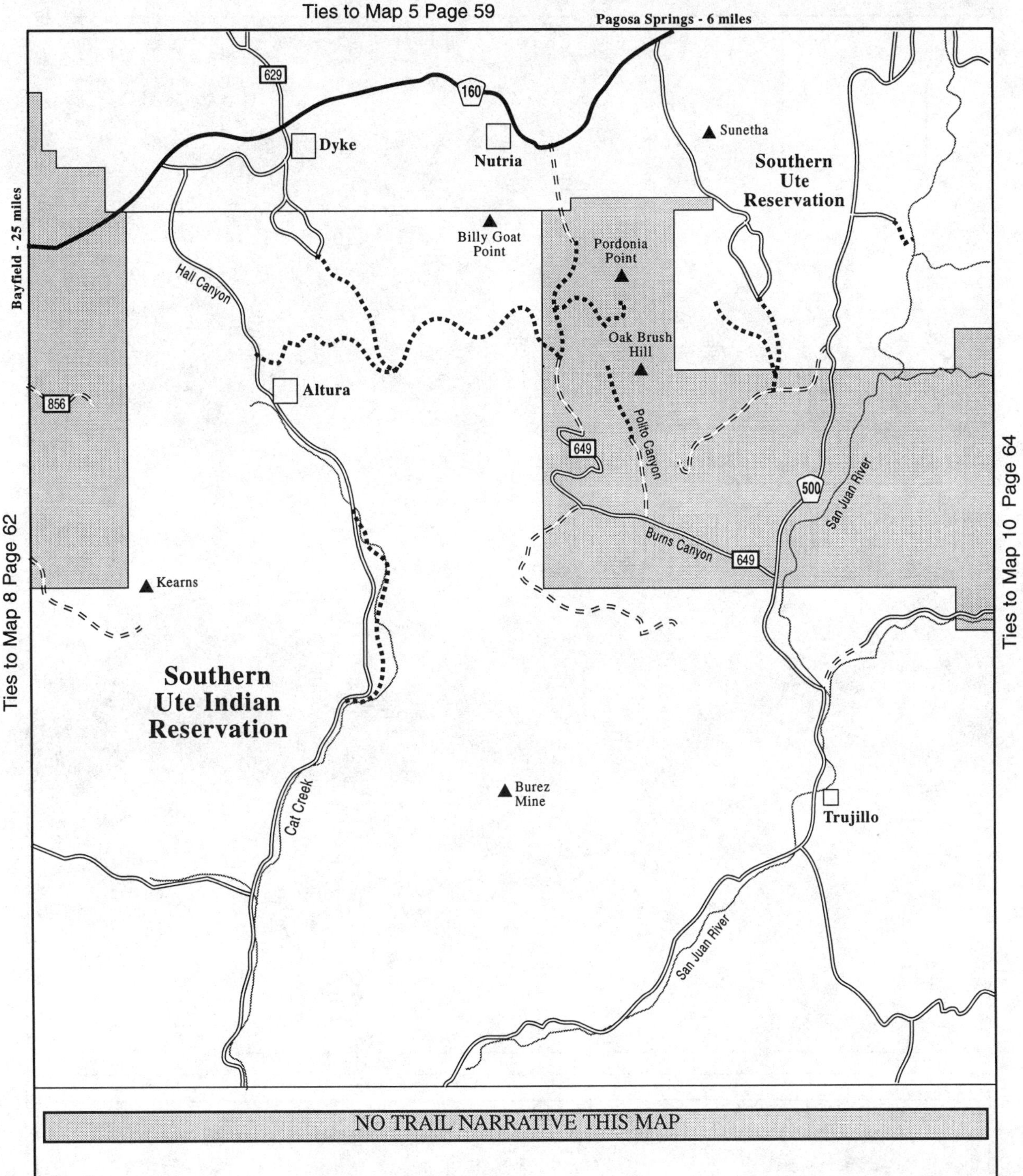

Ties to Map 5 Page 59

Pagosa Springs - 6 miles

[629]
[160]

Dyke

Nutria

▲ Sunetha

**Southern
Ute
Reservation**

Bayfield - 25 miles

Hall Canyon

▲ Billy Goat
Point

Pordonia
Point ▲

Oak Brush
Hill ▲

[856]

Altura

Polito Canyon

[649]

San Juan River

[500]

Ties to Map 8 Page 62

Burns Canyon
[649]

▲ Kearns

**Southern
Ute Indian
Reservation**

Ties to Map 10 Page 64

Cat Creek

▲ Burez
Mine

Trujillo

San Juan River

NO TRAIL NARRATIVE THIS MAP

AREA 2 Map 9

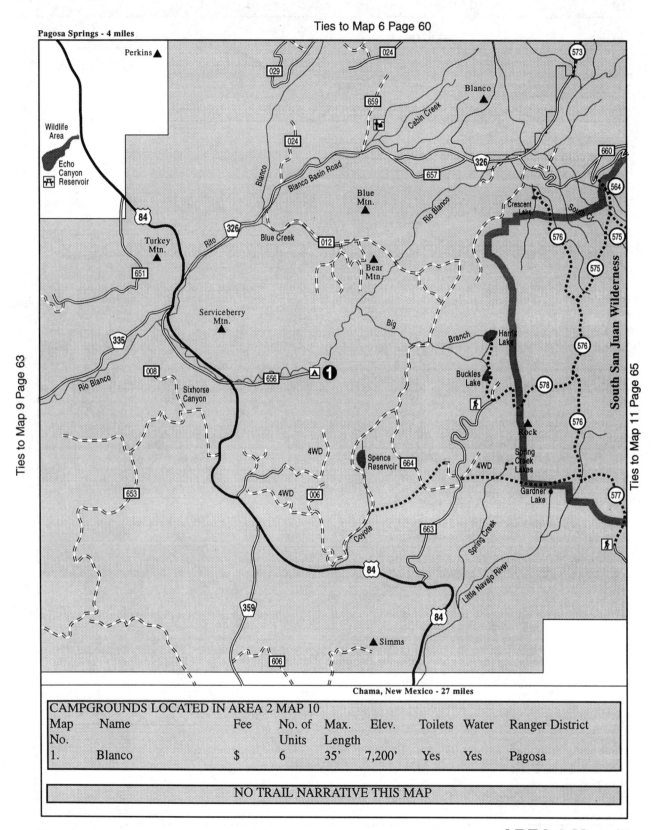

Pagosa Springs - 4 miles

Perkins ▲

029

024

659

Blanco

Cabin Creek

Wildlife
Area

Echo
Canyon
Reservoir

84

024

326

657

573

660

564

Blanco Basin Road

Blanco

Blue
Mtn.

Rio Blanco

Crescent
Lake

South Cr.

576

575

575

Turkey
Mtn.

326

Rito

Blue Creek

012

Bear
Mtn.

651

Serviceberry
Mtn.

Big

Branch

Harris
Lake

Buckles
Lake

576

576

578

335

008

Rio Blanco

656

△ ①

Sixhorse
Canyon

Rock

Spring
Creek
Lakes

577

576

4WD

Spence
Reservoir

664

4WD

Gardner
Lake

653

4WD

006

Coyote

663

Spring Creek

84

Little Navajo River

359

84

606

▲ Simms

Ties to Map 9 Page 63

South San Juan Wilderness

Ties to Map 11 Page 65

CAMPGROUNDS LOCATED IN AREA 2 MAP 10

Map No.	Name	Fee	No. of Units	Max. Length	Elev.	Toilets	Water	Ranger District
1.	Blanco	$	6	35'	7,200'	Yes	Yes	Pagosa

NO TRAIL NARRATIVE THIS MAP

AREA 2 Map 10

NO CAMPGROUNDS LOCATED IN AREA 2 MAP 11

Ties to Map 7 Page 61

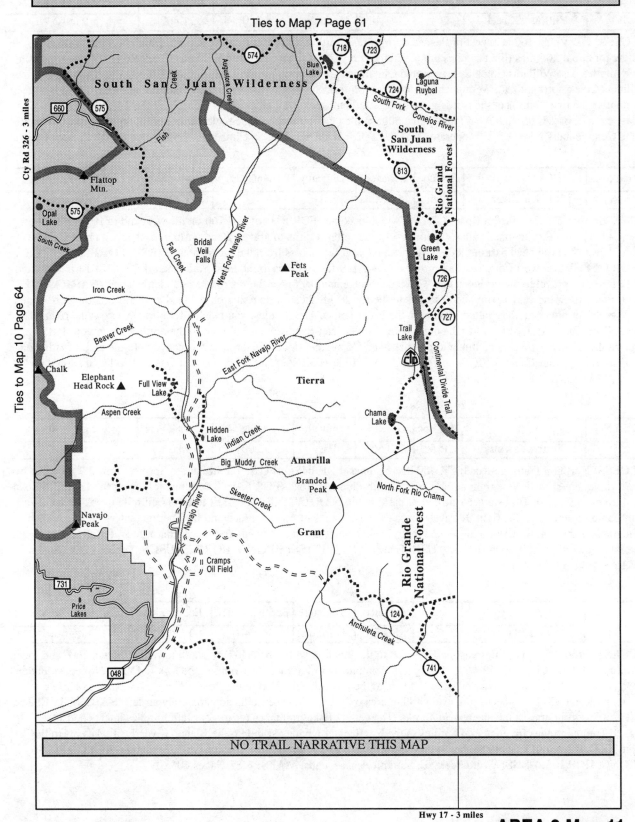

South San Juan Wilderness

574

Blue Lake

718   723

724

Laguna Ruybal

South Fork   Conejos River

South San Juan Wilderness

813

Rio Grand National Forest

Cty Rd 326 - 3 miles

660   575

Fish   Creek

Augustora Creek

Flattop Mtn.

575

Opal Lake

South Creek

Ties to Map 10 Page 64

Fall Creek

Bridal Veil Falls

West Fork Navajo River

Fets Peak

Green Lake

726

Iron Creek

727

Beaver Creek

East Fork Navajo River

Trail Lake

Chalk

Elephant Head Rock   Full View Lake

Tierra

CDT

Continental Divide Trail

Aspen Creek

Hidden Lake   Indian Creek

Chama Lake

Big Muddy Creek

Amarilla

Navajo Peak

Skeeter Creek

Branded Peak

North Fork Rio Chama

Grant

Navajo River

Cramps Oil Field

731

Rio Grande National Forest

Price Lakes

124

Archuleta Creek

048

741

NO TRAIL NARRATIVE THIS MAP

Hwy 17 - 3 miles

AREA 2 Map 11

Trail No.	Trail Name	Map Loc.	Distance	Difficulty	Beginning Elev.	Ending Elev.	Ranger District
587	**Williams Creek**	Q 4	14 Mi.	Moderate	8,360'	11,800'	Pagosa

ACCESS: From U.S. 160 just west of Pagosa Springs turn north onto Piedra Road (CR 600), which becomes FDR 631. Follow for about 26 miles (First 6 miles are paved the rest gravel). Turn right onto Williams Creek Rd (FDR 640) and continue 4 miles past Williams Creek Reservoir and Cimarrona Campground to the trailhead. **NARRATIVE:** The trail follows Williams Creek north into the Weminuche Wilderness. Motorized vehicles and mountain bikes not allowed. There is a steep climb after the first .5 mi., but the rest is gradual. The Indian Creek Trail takes off to the right at the 2.5 mile point. At 3 miles, the trail passes through what looks like a huge walled garden with unique rock formations. It then continues 11 miles to the Continental Divide. **USE:** Moderate. **ACTIVITIES:** HIKING. **USGS:** Cimarrona Peak Quad. **MAP:** 2. Page 66.

Trail No.	Trail Name	Map Loc.	Distance	Difficulty	Beginning Elev.	Ending Elev.	Ranger District
596	**Piedra River**	Q 5	14 Mi.	Easy	7,200'	7,700'	Pagosa

ACCESS: Turn north on Piedra Road (CR 600, which becomes FDR 631) of U.S.160 on the west end of Pagosa Springs. The road is paved for the first 6 miles. Continue 11 miles after it turns to gravel. Head north until the road crosses the Peidra River. The trailhead parking lot is just ahead on your left across from the picnic area. **FIRST FORK SHUTTLE DROP OFF:** Travel west from Pagosa Springs on Hwy 160 for 19 miles to the First Fork Road (FDR 622) Turn north and go 12 miles on gravel to the bridge at end. Cross the bridge and park near the restrooms and picnic tables. **NARRATIVE:** The trail starts on the canyon rim and descends to the river with sheer cliffs rising on both sides. It stays on the north side of the river. River otters can be spotted playing in the water below. At 3.5 miles, you reach a foot bridge across the river. Lower Weminuche Trail forks off to the north here, 2 miles later Sand Creek trail goes to the north. Stay to your left through these intersections to continue to parallel river. Go another 8.5 miles through box canyony to the end at bridge on the Fisrt Fork Hunter Camp. **USE:** Moderate. **ACTIVITIES:** HIKING, HORSES. **USGS:** Oak Brush, Devil Mountain Quads. **MAPS:** 4&5. Pages 58&59.

Trail No.	Trail Name	Map Loc.	Distance	Difficulty	Beginning Elev.	Ending Elev.	Ranger District
579	**Fourmile Falls**	R 5	3.0 Mi.	Easy	9,200'	9,800'	Pagosa

ACCESS: North on Fourmile Road (CR 400 that becomes FDR 645) off U.S. 160 in Pagosa Springs. Travel 9 miles on two wheel drive gravel road. Veer right at the junction with Plumtaw Road (FDR 634) to continue following the Fourmile Road. Go 4 miles to the end. Trail starts at the end of parking lot. **NARRATIVE:** The Trail follows Fourmile Creek up a canyon with Eagle Mountain on the right. It travels through dense stands of aspen, spruce and fir. The trail enters the Weminuche Wilderness 1 mile in, where mtn. bikes and motroized vehicles are not allowed. You reach Fourmile Waterfall 2 miles later. The falls drop 300 feet from cliff above. **USE:** Heavy. **ACTIVITIES:** HIKING, HORSES. **USGS:** Pagosa Peak Quad. **MAP:** 5. Page 59.

Trail No.	Trail Name	Map Loc.	Distance	Difficulty	Beginning Elev.	Ending Elev.	Ranger District
572	**Little Blanco**	T 6	8 Mi.	Difficult	10,040'	12,200"	Pagosa

ACCESS: From Pagosa Springs east on Hwy 160 to the junction with Hwy 84. Turn south towards Chama, NM, and travel .8 mi. to Mill Creek Road (FDR 662) turning left. Continue 6.5 miles and veer right at the fork to follow Nipple Mountain Road (FDR 665). Two miles later take the left fork at the junction with Porcupine Road (FDR 024) to stay on Nipple Mountain Road. Go 9.5 miles to trailhead parking area on your right. The trailhead is on the your left. **NARRATIVE:** The first 2 miles of the trail climb steep switchbacks. The Quartz Lake trail takes off to your left 3.5miles later. The Little Blanco Trail continues for another 4,5 miles to the Continental Divide. At mile 6 the Blanco River Trail takes off to the right. Most of the trail is above timberline so be prepared for exposer to the elements. **USE:** Moderate. **ACTIVITIES:** HIKING, HORSES. **USGS:** Wolf Creek SE, Summit Peaks Quads. **MAPS:** 6&7. Pages 60&61.

Trail Name	No.	Recommended Activities	Page
Animas River	.675	.Hiking, horses, fishing	.22
Bear Creek	.607	.Hiking, horses, mtn. bikes, motorcycles, fishing	.15
Big Bend	.516	.Hiking, fishing	.21
Calico	.649	.Hiking	.17
Calico Nat. Recreation	.208	.Hiking	.7
Cascade Creek	.510	.Hiking	.19
Clear Creek	.550	.Hiking, fishing	.29
Coal Creek	.677	.Hiking	.22
Colorado Trail	.1776	.Hiking, mtn. bikes, horses	.37
Columbine Lake	.509	.Hiking, fishing	.11
Continental Divide Trail	.813	.Hiking	.11
Continental Divide Trail	.813	.Hiking, horses	.47
Corral Draw	.521	.Hiking	.21
Crater Lake	.623	.Hiking	.25
Cross Mountain	.637	.Hiking, horses	.9
Cunningham Gulch	.502	.Hiking, fishing	.10
Deer Creek	.678	.Hiking, fishing	.23
Dutch Creek	.516	.Hiking	.20
East Creek	.535	.Hiking, horses	.33
East Fork	.638	.Hiking, fishing, mtn. bikes	.9
Elbert Creek	.512	.Hiking	.20
Elk Creek	.503	.Hiking	.25
Engine Creek	.657	.Hiking	.22
Engineer Mountain	.508	.Hiking	.19
First Fork	.727	.Hiking	.30
Fish Creek	.647	.Hiking, horses, fishing	.7
Fourmile Falls	.579	.Hiking, horses	.66
Geyser Spring	.648	.Hiking	.7
Gold Run	.618	.Hiking, horses, motorcycles	.16
Goulding Creek	.517	.Hiking	.20
Graysill	.506	.Hiking	.19
Haflin Creek	.557	.Hiking	.30
Hermosa Creek	.514	.Hiking, horses, mtn bikes, motorcycles, ATV's	.29
Highland Mary	.606	.Hiking, fishing	.11
Highline Loop Nat. Rec.	.607	.Hiking, horses	.27
Horse Creek	.626	.Hiking, horses	.16
Ice Lake	.505	.Hiking	.9
Johnny Bull	.639	.Hiking, horses	.16
Jones Creek	.518	.Hiking, horses	.29
Kennebec Pass-Junction Crk. Rd.	.553	.Hiking, mtn bikes	.27
Lake Eileen	.668	.Hiking	.34
Little Blanco	.572	.Hiking, horses	.66
Little Elk Creek	.515	.Hiking	.20
Loading Pen	.738	.Hiking	.13
Lost Lake	.663	.Hiking, fishing	.33
Molas	.665	.Hiking	.25
Needle Crk/Chicago Basin	.504	.Hiking	.25
North Canyon	.656	.Hiking, mtn bikes, horses	.33
Pass Creek	.500	.Hiking	.19
Piedra River	.596	.Hiking, horses	.66
Priest Gulch	.645	.Hiking, horses	.17
Red Creek	.726	.Hiking	.34
Rico-Silverton	.507	.Hiking	.9
Rough Canyon	.435	.Hiking, horses, fishing	.15
Ryman Creek	.735	.Hiking, horses, fishing	.17
Salt Creek	.559	.Hiking	.21
Sharkstooth	.607c	.Hiking, horses	.27
Shearer Creek	.558	.Hiking, horses	.33
Sliderock	.622	.Hiking	.30
South Fork	.549	.Hiking, fishing	.29
Spud Lake	.661	.Hiking, fishing	.22
Tenderfoot	.644	.Hiking, fishing	.16
Twin Spring	.739	.Hiking	.17
West Lime Creek	.679	.Hiking	.23
West Mancos	.565	.Hiking, mtn. bikes, horses	.27
Whitehead	.674	.Hiking	.11
Wildcat	.207	.Hiking, horses	.15
Williams Creek	.587	.Hiking	.66
Youngs Canyon	.546	.Hiking	.33

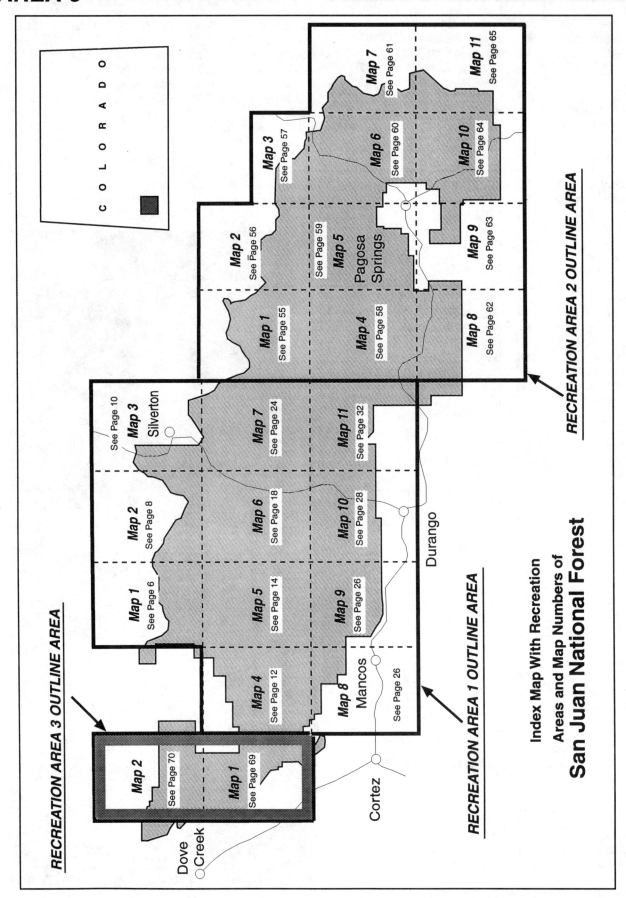

Index Map With Recreation
Areas and Map Numbers of
San Juan National Forest

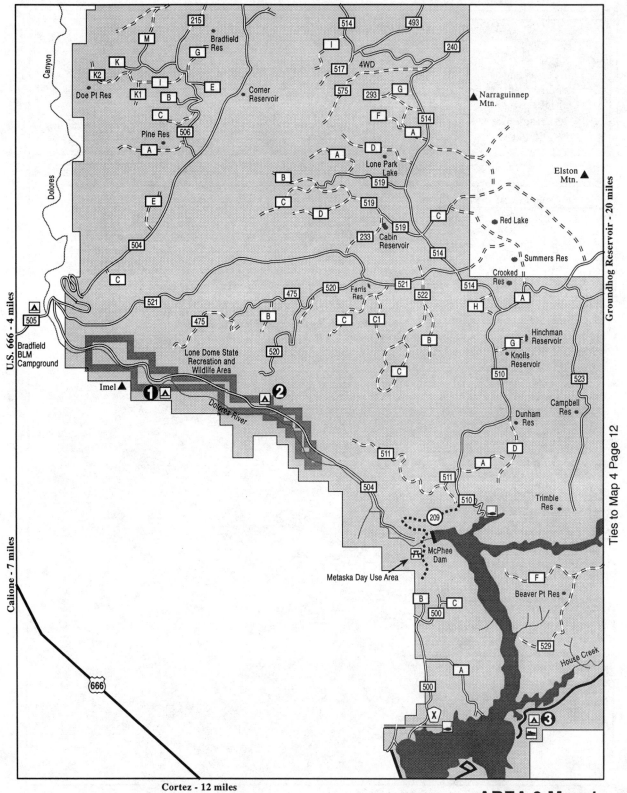

NO TRAIL NARRATIVES AREA 3

Ties to Map 2 Page 70

AREA 3 Map 1

Map No.	Name	Fee	No. of Units	Max. Length	Elev.	Toilets	Water	Ranger District
	CAMPGROUNDS LOCATED IN AREA 3 MAP 1 (Page 69)							
1.	Cabin Canyon	$	11	45'	6,500'	Yes	Yes	Dolores/Mancos
2.	Ferris Canyon	$	6	45'	6,500'	Yes	Yes	Dolores/Mancos
3.	McPhee RA							
	A. House Creek & Group	$	55	50'	7,100'	Yes	Yes	Dolores/Mancos
	B. McPhee	$	73	50'	7,100'	Yes	Yes	Dolores/Mancos

NO CAMPGROUNDS LOCATED IN AREA 3 MAP 2

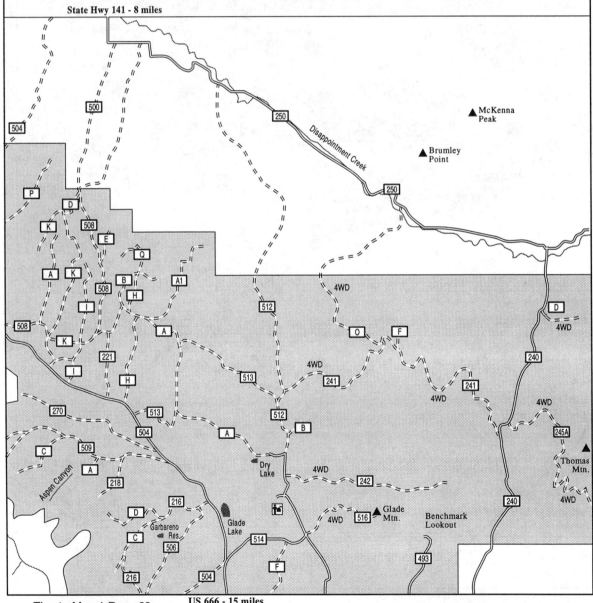

Ties to Map 1 Page 69

US 666 - 15 miles

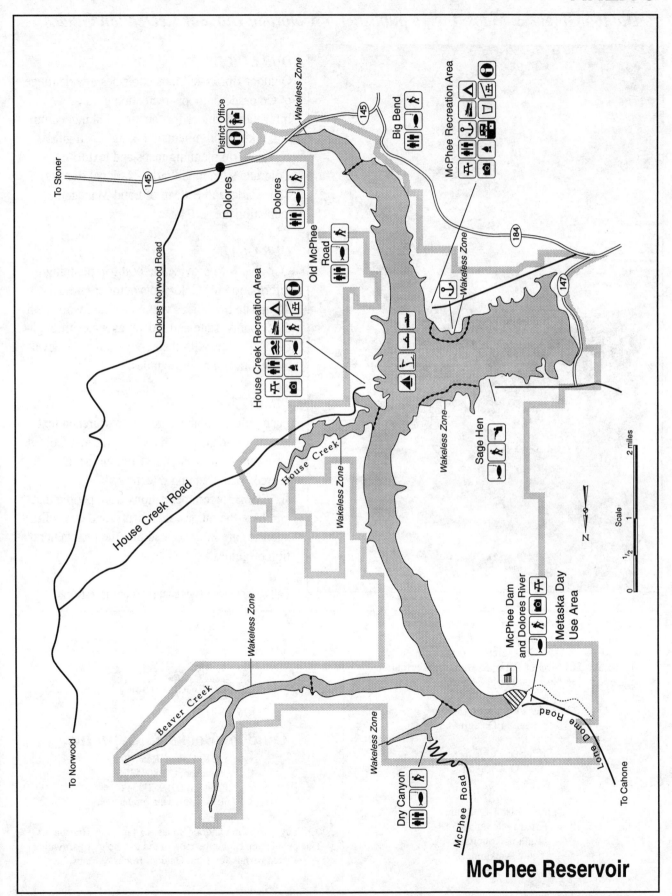

McPhee Reservoir

Outdoor Books & Maps, Inc., a publisher of Colorado Outdoor Recreation Guides

OUR GUIDES

Outdoor Books & Maps publishes a wide range of Colorado outdoor recreation guides. Informative guides for boating, fishing, camping, hiking and mountain biking. Each guide contains trip planning maps and text to Colorado's National Parks, National Forests, State Parks and Bureau of Land Management recreation areas.

OUR GOAL

Our goal is to provide the best trip planning information for Colorado's outdoor enthusiasts. To update the guides we contact and work with city, county, state and federal agencies throughout Colorado. With their cooperation we develop accurate and fresh guides.

QUALITY

Each guide contains carefully researched text and maps updated with each reprinting. Current information not avaiable from any single source. Trip planning directories!
We listen! Recommendations from people that purchase the guides, merchants, and controlling agencies are incorporated into the next generation of guides.

Call or write with your comments or suggestions.

Acknowledgements
U.S. Forest Service
U.S. Geological Survey
Colorado Division of Parks & Outdoor Recreation
National Parks Service

Graphics and Cover Design
Grasman Design

Editor & Publisher
Jack O. Olofson

Staff
Jackie Ellison-Quintana
Kristin Alexander
Linnea Roberts
Cindi Alexander
Dody Olofson

Outdoor Books & Maps, Inc.
11270 County Road 49
Hudson, CO 80642
Phone: (800) 952-5342
E Mail: obm@iguana.ruralnet.net